Mac Cooper
A Biography

Joan Smith.
With best wishes.,

John Craven. 18. Sept 2001

Mac Cooper
A Biography

John Craven

The Pentland Press Limited
Edinburgh • Cambridge • Durham • USA

© John Craven 2000

First published in 2000 by
The Pentland Press Ltd.
1 Hutton Close
South Church
Bishop Auckland
Durham

British Library Cataloguing in Publication Data.
A Catalogue record for this book is available
from the British Library.

ISBN 1 85821 807 1

Typeset by CBS, Martlesham Heath, Ipswich, Suffolk
Printed and bound by Antony Rowe, Chippenham, Wiltshire

Dedication

To the three 'whirlwinds'
Barbara, Diana and Cynthia.

With love

CONTENTS

FOREWORD

By Professor John Prescott
Principal of Wye College, University of London

This biography will inform and inspire with its humanity and its tale of enterprise and commitment. It tells of a life spanning the greater part of the twentieth century during which Professor Mac Cooper became one of the outstanding agriculturists in the United Kingdom. He was both a visionary and a leader who challenged conventional thinking. At the end of his career, both here in the UK and in New Zealand, he received just recognition for his outstanding contribution to agricultural education and to the extension of research into new farm practice.

I write this foreword as one of Mac's former colleagues who was privileged to work within his team at Newcastle and Cockle Park in the 1960s. He gave his young staff extraordinary scope and my subsequent life and career owes much to his influence. He was both an inspiring lecturer and a gifted writer; listened to and read with rapt attention by audiences who could apply his message to the practical realities of the day. He put his ideas across with such clarity and resonance that they stuck in the mind of his listeners and readers. In short, he was an outstanding communicator.

Readers of this biography will be indebted to John Craven for providing such an insight into the life and career of Mac Cooper. John himself, as Head of the Milk Marketing Board's Farm Services Division, had a considerable influence on the UK dairy sector and he writes with authority about the developments in the agricultural industry that he has known first hand during the past thirty years. As a former postgraduate student of Mac's at Newcastle and also his son-in-law, John has produced a narrative that is as fascinating as it is readable. Like Mac's own books, once taken up it is difficult to put down. He draws on a wealth of sources, including unpublished correspondence, university archives and war records as well as Mac's own publications. He is frank and direct in his writing. He does not shy away from sensitive issues, including personality clashes. He deals openly and fairly with the difficulties that arose at different stages in Mac's career and he brings particular sensitivity and insight to the interaction between his personal and professional life.

In spite of the inevitable tensions arising from such a demanding schedule of teaching and public service, the inherent strength of Mac's marriage and family life clearly provided the bedrock on which his dynamic career was founded. He never lost touch with his New Zealand roots and he shared great joy with his wife Hilary in their family. This biography does justice to a remarkable man who was a leader in agricultural development and who also had a profound influence on so many individual students, colleagues and farmers. I trust that you will enjoy this book as much as I have.

INTRODUCTION

In the foreword to a book which Mac Cooper began to write towards the end of his life he states: 'I am a somewhat reluctant auto-biographer, because in a previous attempt some five years ago, I gave up even before I got to Oxford because of a growing distaste of the first person.' He decided to call his book *Recollections of a Rather Rural Dean* and chose Shakespeare's famous quotation from *As You Like It* upon which to base the seven ages of his life. He only got as far as the 'lover' and the woeful ballad was not to a beautiful girl, but to a shapely Tamworth gilt that he had reared when looking after the pig unit at Massey Agricultural College. Sadly, his second attempt fell short of Oxford as well and we can only guess at what he might have written had he come to the soldier, 'full of strange oaths' or the justice 'full of wise saws and modern instances'.

Mac Cooper, apart from the first and last decades, lived through virtually the whole of the twentieth century. He was born at about the same time as Edward VII died: a time when the sun never set on the British Empire and even America regarded her as the supreme world power. Every generation will feel that they have lived through a period of massive change and Mac's must rank as extreme as most. In the foreword to his book, he reflected on some of these changes.

> Consider the political and structural upheaval caused by two world wars, the re-building of Europe and its expansion into an Economic Community, and the unexpected collapse of communism in Russia. Think also of what has happened to travel and communication since the early turn of the century when the Wright brothers and Marconi, working in their different spheres initiated a conquest of the air, which brought the ends of the earth into close spoken and visual contact. In 1934 it took me five weeks to travel from Auckland to Tilbury and at least twelve days to get a reply to letters posted in either Britain or New Zealand. Now you can be at your destination within 24 hours or watch the British Lions facing up to the All Blacks on satellite TV even if it is 3 am in the morning. With fax or E-mail, communication is instantaneous and telephone conversations are often clearer than those with the next door neighbours.

Possibly the most dramatic changes in Western societies have been in working practices, living standards and life-style. A farm worker in Britain prior to 1934 worked a 12-hour day and a six-day week without any holiday entitlement. Often in the short winter days the first task of the ordinary labourer would be hand spreading of farmyard manure which had previously been dumped in the field. This was a job he could do from memory or with the faint light of a hurricane lamp. Horsepower was literally what it means, not the size of a tractor engine. Factory workers and miners, because they were better organised, had some relief from the long days of incessant toil that was expected of them, but even as recently as the 30s it was only the privileged few who travelled abroad for pleasure. Now this is commonplace and modern hotels have transformed small fishing ports along the Mediterranean coast, where it is English and German rather than Spanish or French that is commonly heard in the streets for most of the year. Fishing still survives for it is a source of raw material for an essential home comfort — fish and chips — providing sustenance for those who have not yet lost their prejudices against foreign food. But the new age travellers are learning fast and back home they will frequently treat themselves to restaurant meals washed down with a carafe of wine. The average Briton is a very changed animal from what he was 50 years ago when a week in Blackpool, Brighton or Butlins was indeed a high adventure.

Compared with the past, and with present conditions in many developing countries, we certainly live in an affluent society and for most people the slogan which Harold Macmillan once used to such good electoral effect, 'You never had it so good,' has the ring of truth.

But politicians have had only a marginal effect on these profound changes in material standards of living. They have been no more than manipulators and sometimes meddlers in a process that was initiated firstly by a technological revolution which greatly increased the availability of goods and services and secondly by the steady breakdown of the class society where everyone knew their place. The arrival of a Labour Government after the second world war with the commitment and majority to create a welfare state, reduce unemployment, and provide easier access to higher education was a watershed of huge significance. Advances in technology were responsible for what might be best described as a chain reaction. They contributed to higher productivity and increased the range of consumer goods as well as lowering unit costs of production in real terms. Higher productivity brought with it improved incomes and increased demand. This in turn led to growth in jobs, even to the point where immigration from the new Commonwealth countries was encouraged, particularly for repetitive and menial tasks.

Neither have women missed out on this chapter of change. From the dramatic days of the suffragettes and work opportunities created during the war years, they now hold jobs throughout industry, commerce and the professions, admittedly not always with rewards matching those of their male counterparts. Generally women are no longer slaves to the kitchen sink and often they are significant contributors to family incomes. But with this emancipation has come the inevitable trend towards increasing independence and the consequent cultural change to family life and values. British institutions such as Sunday lunch where the whole family would gather to eat and discuss the events of the week seem now to be the exception rather than the rule. Indeed, some would argue, a tray on the lap in front of the TV with a supermarket pre-prepared meal is more the norm — with conversation minimal in case the most recent soap revelation is missed. Now with increased awareness and availability to contraception, women are more in charge of their own destiny. To have children outside marriage is no longer a social stigma and one-parent families are now commonplace. All this is a far cry from family life and the social values which prevailed in Britain and even more so in New Zealand at the turn of the century.

The majority of people in Britain and the old Commonwealth countries had no more than an elementary education and only a privileged few went past secondary school to full time University education. Even as recently as 1940 there were only eighteen universities in the United Kingdom. Now with the elevated status which degree courses confer, Universities, Polytechnics, and Colleges of Further Education seem to exist in every city in the land. Neither is there a shortage of applicants (except where there has been over proliferation of courses) for the wider availabiity of secondary education has been encouraged with grant aid for students who have achieved the necessary entry qualifications. More recently, however, the funding of fees and the concept of treating grants as repayable loans, have discouraged some applicants. Nevertheless it remains true that increasingly the leaders of industry are now the products of state education and redbrick universities.

Reflecting on increased educational opportunities over those early in the century, Sir John Hammond said that by far the brightest member of his class at the village school was the son of a cobbler. He had had to leave school aged twelve, because the family needed the few shillings he could earn at that tender age. John, a farmer's son, was able with family help and scholarships, to progress through Cambridge to a distinguished postgraduate career. By middle age he was an internationally acclaimed animal physiologist and a Fellow of the Royal Society. He was also appointed a knight of the realm in recognition of his distinguished services. What would have been the heights attained by the cobbler's son if he had been born two generations later?

In my time at Newcastle there was a student, one of a family of four who all had achieved first class honours degrees. Their father was a labourer at a waterworks in a remote Pennine village. But such a background is no longer a handicap where there is inherent academic ability that is nurtured by dedicated teaching, first at primary school and then later at the secondary level. Society is so much richer in consequence because most inherently able boys and girls are now being given chances that were denied earlier generations. This is a matter on which I have strong convictions. Apart from my own situation coming from a background of family farming in New Zealand when university education was exceptional, my career has been mainly concerned with young men and women during their most formative years while at university. More than this I have been working in a discipline that has been a life-long interest and also a way of life for my family. Generally people who read agriculture do so not because they see a degree as a stepping stone to great monetary rewards. They choose that course because of their love of farming and their interest in land and livestock.

Many come from family farms and are the product of the smaller secondary schools that do not have the resources of large schools in preparing students for university. Often they would come up with no better than a couple of Ds at A level, but at interview one could see that they were the right sort of material and so this was the modest target they were given in order to qualify for entry. It was a joy to observe how most of them developed in a university environment and after they graduated took up responsible positions, not only in Britain but also abroad. University teaching is potentially a very rewarding occupation and I will always be grateful that it has been related to my other great interest which is practical farming.

If I have regrets at the end of a working lifetime, possibly it is because I have been unable to make more than a trivial contribution to the advancement of agriculture in developing countries. I have visited Africa and South America and I have given advice on the organisation of both higher agricultural education and research. In addition many postgraduates from these countries, of varying quality and degree of dedication, have passed through my hands. Sadly there are too many that are more interested in personal advancement than in helping to improve agriculture in their own home countries. This is hardly surprising with young people who have come from very primitive backgrounds — homes where there is no literacy and with parents who are no more than subsistence farmers. It is asking a lot of such people to have the same attitudes as those for whom service to the community is implicit in the planning of their careers, particularly where they go into applied fields such as medicine, engineering and agriculture. Unquestionably the greatest challenge now facing my profession, the politicians and the administrators as well, is the betterment of agriculture in third world countries where the spectre of famine can too often become a reality. There is still much to be done to make the world a better and safer place in which to live, but the example of British agriculture over the past fifty years gives one hope. In the early nineteen thirties, it was a moribund industry but today it compares favourably with the agriculture of any western country because advances in technology and related inputs including capital, have been purposefully applied. It will always be a great source of personal satisfaction to have been, in a small measure, a contributor to this spectacular change.

That there continues to be change and fluctuating fortunes, some may say with increasing pace, is self-evident. Even so, Mac Cooper would probably have been appalled to see the degree to which the farming industry has declined in terms of image in the space of the eleven years since his death. For decades, farmers have enjoyed a largely sympathetic public. During the two world wars they were respected for their contribution to the task of keeping the country supplied with essential food. They were urged to increase productivity to get rid of rationing, and Government invested heavily in financial incentives to modernise British farming in order to reduce reliance on imports. But with our entry into Europe, the problem of food surpluses, coupled with the cost of taxpayer support, began to erode public confidence. Quite rightly, people could not reconcile the complexities of the market on the one hand and the obscene TV pictures of starvation in various parts of the world on the other, especially as they seemed to be projected into their living rooms each evening.

Farming at the start of the second millennium in Britain is in severe recession. There have always been cycles, due either to the weather or to depressed prices, but farmers are used to that and most plan their operations to tide them over the lean times. But in addition to the constant barrage of media hype from the extremist lobby, there also appears to be a more worrying trend following in the wake of recent food scares. It seems that farmers and scientists are no longer to be trusted. A significant proportion of the general public even argue that there should be no experimental evaluation to assess the merits of genetically modified crops, despite the potential impact that these might have on reducing starvation in some of the world's poorer countries. Few would disagree that such new technology needs the utmost scrutiny, but it seems, if some parts of the media are to be believed, that the subject is far too important to be left in the hands of scientific research, especially if it is funded by commercial self-interest.

But whatever the outcome of such debates there remains for European agriculture the fundamental problem of how to break out of the straightjacket of subsidies. This was a subject close to Mac Cooper's heart, when in the fifties he was urging Government to recognise that such a policy was bound to be flawed in the long term. It is salutary that farm subsidies were removed entirely from New Zealand agriculture in the 1980s, so it is possible for farming to survive without them. But the political will to implement such a sea change within an expanding European Community seems as far distant as ever.

Farming of course will recover. There are many thousands of well run family businesses which, despite all the pressure, remain profitable and who have the resilience to withstand all that is thrown at them. They will, however, have to look increasingly at non-farming ventures to supplement income, especially on the smaller holdings. There are many others, maybe not so small, who will have to alter their strategy fundamentally if they are to remain in the industry.

Mac's other great interest, agricultural education, has also changed almost beyond recognition since the mid-sixties when he was at the peak of his career. There are now courses available in every specialist interest; teaching methods have been revolutionised through the advances in computer technology and the general degree in agriculture with its wide mix of science and practice has now become a minority interest. Whether he would have approved of these changes is debatable. He was a realist but he also believed passionately in the rounded view so that students could gain some perspective of this complex industry and the way in which it worked. This was largely the reason why he was able to command such attention when he chose to speak on agricultural topics of the day. Apart from having a natural talent in the art of communication, he had the intellect and vision to see the wood as well as the trees. Leaders of our academic institutions nowadays are inevitably more specialised as a result of their own training and the industry has become such a political football that sadly their contribution to the national debate seems not to carry much influence.

Martin Johnson, the cricket correspondent of the *Daily Telegraph*, when reporting on the defeat of New Zealand by Pakistan in the World Cup semi-final of 1999, wrote:

New Zealanders take mild umbrage at their team being labelled dull, as they do when their country is similarly described. However, it is hard to escape once you have been stereotyped. I recall reading a foreword in a New Zealand travel book (probably written by an Australian) which read 'All the people with get up and go, have got up and gone.'

Of course this is utter nonsense, but Mac Cooper, for whatever reason, decided to make Britain, rather than his home country, the focus of his attention. For many of his students and its farmers this seems indeed to have been a fortunate choice.

ACKNOWLEDGEMENTS

Readers will soon realise that this biography of Mac Cooper is not a scholarly work, but an attempt to document a life, based largely on the personal observations of a cross-section of people who knew him: colleagues, students, friends and family.

As one might expect, views of events and personalities differ, but there has been a universal enthusiasm for the project from everyone with whom I have made contact. He clearly influenced many lives.

There have been two major sources of information which have been used in this work. First, extracts from a draft of Mac's own attempt at an autobiography, which he started to write towards the end of his life, and second, direct quotes from many of the personal letters which he wrote to members of his family. These not only give a new slant on some of the principal events in his career, but are also an insight into his private life and sense of values. The most frequent recipient of these family letters was one of Mac's sisters, Thelma Porter, who lives in Hastings, New Zealand. Nearing her own centenary, I am most indebted to her for allowing me to use this material and also for her reminiscences of a much-loved youngest brother. Her daughters, Sally Combly, Billie Campbell and Jan Porter, together with her niece, Bev Muggeridge, have also helped research the history of the family and followed up many of the leads which I established during my own research.

Also in New Zealand, Professor Ian Campbell kindly showed me round the campus at Massey and provided much help in accessing records from their archives. Also my thanks to Lance McEldowney, Editor of *New Zealand Dairy Exporter* for his help and in particular his article in the May 1999 edition of the magazine outlining the project. I am grateful to many others in that country, including Sir Alan Stewart, Rex Patchell, Mat Sanders, Al Wray, Clive Dalton, Peter Mortimor, Neil Worker, Gordon McLaughlan, Bill Brien and Margaret Smith for their generous help and time. My thanks as well to Glen Whittaker, an old friend and sparring partner, formerly General Manager of the Livestock Improvement Corporation, who organised and scheduled my trip to New Zealand.

I appreciate the agreement of the Warden of the Rhodes Scholarship Trust, Dr John Rowett, to allow me to use material from Mac Cooper's personal file, which is stored in Rhodes House at Oxford. James McNeish, the celebrated

New Zealand author, whom I met in Oxford, also helped me considerably with his account of the background leading to Mac Cooper's Nuffield award.

At Wye, I received exceptional assistance from Professor John Prescott, Principal of the College, and from his library staff who allowed me access to their extensive archives. John has also been kind enough to agree to write a forward to this book. I met many people at Wye, amongst whom Canada Davies, Ian Reid, Donald Sykes, Bill Holmes and Louis Wain contributed anecdotes as well as accounts of their contacts with Mac Cooper when he was a colleague during his time at the College. There were many others to whom I spoke about the project and to all of them I am most grateful for their help and encouragement.

At Newcastle, there is less documented information upon which to base research. However, Wilma Lister, the Honorary Secretary of the University Agricultural Society, worked endlessly on my behalf to locate both people and material which might be relevant to the subject and I thank her most sincerely for her efforts. My thanks also to some of Mac's former colleagues at Newcastle for their invaluable input, especially Jim Merridew, Bill Weeks, Graham Ross, Fred Blackburn, Bob Thomas, Malcolm Willis, Maurice Bichard and Gordon Dickson. Others, as former students at Newcastle and now members of staff, generously offered their opinions and recollections.

There were many other rich sources of information in the North-east, including Ted Pears at Cockle Park, John and Liz Craigs, John Moffitt, Dr Ken Baker, Jay Bosanquet and others to whom I offer my thanks.

I am also indebted to Mike Finch, editor of the *Farmers' Guardian*, for allowing me access to their microfiche copies of previous editions of that newspaper.

This appreciation so far has referred to the help I have received from people by talking to them personally. In addition I received a great deal of correspondence from various parts of the world when it became known that I was engaged in this project. I have quoted extensively from this correspondence in the manuscript and I am most grateful to all of those who wrote to me.

Finally, this book would not have been written had it not been for the encouragement offered by my wife Barbara. She and her twin sisters, Diana and Cynthia, provided constructive and tactful criticism throughout and it is with pleasure that I dedicate the book to them.

All the world's a stage
And all the men and women merely players:
They have their exits and their entrances;
And one man in his time plays many parts,
His acts being seven ages. At first the infant,
Mewling and puking in the nurse's arms.
And then the whining schoolboy, with his satchel,
And shining morning face, creeping like a snail
Unwillingly to school. And then the lover
Sighing like furnace, with a woeful ballad
Made to his mistress' eye-brow. Then a soldier,
Full of strange oaths, and bearded like the pard,
Jealous in honour, sudden and quick in quarrel,
Seeking the bubble reputation
Even in the cannon's mouth. And then the justice,
In fair round belly with good capon lin'd,
With eyes severe, and beard of formal cut,
Full of wise saws and modern instances;
And so he plays his part. The sixth age shifts
Into the lean and slipper'd pantaloon,
With spectacles on nose and pouch on side,
With youthful hose, well sav'd a world too wide
For his shrunk shank; and his big manly voice,
Turning again towards childish treble, pipes
And whistles in his sound. Last scene of all,
That ends this strange eventful history,
Is second childishness, and mere oblivion,
Sans teeth, sans eyes, sans taste, sans everything.

William Shakespeare (*As You Like It*)

1

FAMILY BACKGROUND

Malcolm McGregor Cooper, known as Mac, was born on 17 August 1910 in Havelock North, Hawkes Bay, New Zealand. He was the youngest of seven children, five brothers and two sisters. As the baby of the family, he was much adored and according to his surviving sister Thelma, now well into her nineties, 'he was the apple of everyone's eye'. She remembers with affection his 'kiss curl' which the females of the family used to encourage with spit!

The children were third generation New Zealanders; their grandfather James Laurence Cooper, the tenth child in a family of twelve (including three sets of twins) emigrated from Haxey in Lincolnshire at the age of twenty-two. He was from a farming background but there is no record of why he decided to set out on such a hazardous adventure involving a four-month sea crossing and virtually no knowledge of what he would find at the other end.

Conditions in England in the mid-nineteenth century were pretty grim, especially for those working on the land. The industrial revolution was in full swing and the resulting flow of people from the rural areas into the towns and cities was accelerating. It was also a time of enforced eviction of smallholders who had long held undocumented legal rights to land. The Enclosure Acts forced many families from their properties and the men folk were either left as farm labourers or resigned to trudge into the towns looking for work. For the more adventurous, emigration was an option, with New Zealand appealing to many of those with some experience of working on the land. So perhaps it was with this background that James Cooper arrived in Clive, midway between Hastings and Napier, aboard the *Warspirit* on 18 June 1863. He soon found work as a farm labourer on the Te Mata estate owned by the Chambers family at nearby Havelock North.

The principal towns in this area of Hawkes Bay were named after English generals who had been prominent in suppressing the Indian Mutiny of 1858. General Sir Henry Havelock was known as the saviour of Lucknow and in fact there were two settlements which bore his name, the second in the South Island on Pelorous Sound. Some years later a row developed between the two as to which should alter its name to avoid confusion, especially for the postal services. Neither community would give way and thus, the addition of North became a permanent post-script. At this time the 65th Regiment of Foot was stationed in Napier, having been transferred from India, much to the delight of the earlier settlers. It appeared that some of the original purchasers of property in the area were officers from the regiment. The Indian connection continued in subsequent years with such names as Lucknow Road and Simla Avenue being added to the expanding urban area.

Much of this local history is taken from S.W. Grant's book, *Havelock North – From Village to Borough, 1860 – 1952*. In it he describes Te Mata as having been purchased by John Chambers in 1854 and extending to some 6000 acres. The estate employed a substantial number of people including shepherds, ploughmen, labourers, tree planters and extra hands for shearing and haymaking. Chambers came originally from Derbyshire, arriving in New Zealand via South Australia, and it is clear that he had a major influence in the area and its subsequent development. Grant quotes from the *Hawkes Bay Herald*, which describes the inaugural show of the local Agricultural Society, held on 14 August 1863, only two months after James Cooper had arrived in the country. 'The highlight was a ploughing match between two horse drawn teams entered by Mr Chambers and Mr Russell. Ploughing for the former was J. Cooper and the latter G. Peebles. Peebles was compelled to withdraw, as his plough, borrowed for the occasion, was not properly set. Cooper made pretty and even work and was proclaimed the winner.'

In 1865 James married Mary Letitia Taylor who had emigrated with her parents from Offally in Ireland. She was also employed on the estate as a maidservant. They were both twenty-four. Large families were the norm in those days and the couple had nine children, three of whom died at a very young age. Laurence Taylor Cooper, or Larry as he was called, was the fifth child born in 1872 and he subsequently became the father of the family within which Mac Cooper was the youngest child. Mary Letitia died in 1878 at the young age of thirty-seven, leaving James to bring up their six surviving children. By this time James had left the Te Mata estate and taken up work as a coachman, possibly self-employed. One of his fares was the transport of the Vaughan family, recently arrived in Clive from Gloucestershire in England. It is reputed that by the time James had driven them the few miles to Hastings, he had proposed to one of the daughters, Margarette Loughlin Vaughan. Apparently unattached women were very scarce in those days! She accepted, although whether at the time she knew of the children she was also taking on, no one knows.

They were married in September 1879, less than twelve months after the death of James' first wife. Over the next fifteen years they had a further eight children, four girls and four boys, the youngest of whom, Jack, was killed in

France during the First World War. So all in all, James sired seventeen children. By the time the youngest was born, the eldest would have been twenty-eight.

Although at one stage James acquired a small farm, he lost it during one of the many downturns in wool prices. According to Thelma, he was not considered a good provider and it was evident that much of the welfare of the family was in the hands of Margarette, described by her granddaughter as a 'wonderful little woman'. She took all her step-children under her care and ordered the local Catholic priest out of the house so that she could set to work to bring them up as good Anglicans. Things could not have been easy for the Cooper family in those days. Even as a ten-year-old Larry set out to look for part time work outside school hours.

After he explained his determination to contribute something to the rapidly expanding household, W.A. Couper, a local landowner at The Mount Erin Station, gave him a job and at the same time provided a steady guiding hand in future years. The two remained lifelong friends. Further education for most was considered a luxury and Larry did not stay any longer at school than he had to. Before he could realise his ambition to farm on his own account, however, it was necessary to turn his hand to a wide variety of jobs to help support the expanding family. He married Sarah Ann McGregor in 1897 at the age of twenty-five; she was one year older. Sarah Ann was also a second generation New Zealander; her father, Malcolm McGregor (after whom Mac

The Cooper Clan, circa 1914.
Left to right – Thelma, Larry (father), Tom, Mac, Dick, Jim, Madge, Sarah (mother) and Laurie.

3

was named, had emigrated from Scotland in the 1830s. He arrived in the South Island where many of his countrymen had already established themselves and eventually purchased the Longliffe Station in Otago. The family lived there for two years and then lost everything due to an outbreak of sheep scab, which also devasted the livelihoods of many of his neighbours. After managing farms for other people, the family eventually settled in Kakanui on a small block of land where they grew oats and barley.

Malcolm McGregor was a proud man and unlike most other immigrants had arrived with sufficient money to establish himself. It must have been a severe blow to his pride to lose the majority of his capital. Even so, he would not allow any of his daughters to go out to work with the result that Sarah Ann eventually left home without her father's blessing, to seek her fortune in the North Island. She arrived in Havelock North with a sound knowledge of housekeeping, but no other training. However she soon found herself a job as a cook and within a short space of time had joined the Cooper clan.

The newly weds set up home in a cottage on Riverbank which Larry had built with the help of family and friends. In addition to his various manual jobs, he also began to lease small sections of land from neighbours as a start to his farming aspirations. In keeping with the Cooper tradition, their family began to arrive in quick succession, commencing with James (Jim) Laurence in 1898 and Tom Alexander the following year. The two girls, Margaretta Isabel (Madge)

Mac Cooper aged two – the youngest of seven. 1912.
Left to right: Jim, Tom, Madge, Thelma, Laurie, Dick, Mac.

4

and Thelma Anne were born in 1901 and 1903 respectively, followed by two more sons, Laurence Neil (Laurie) in 1906 and Richard Seddon (Dick) two years later. At this stage Riverbank cottage was bursting at the seams and the family moved to a slightly larger house a short distance away. It was here in 1910 that Mac was born. They were a close knit family and the house was fondly called the 'Cosy Ship'. It remained in the family until the mid-1930s, well after Sarah Ann's death. In fact the house was still standing in April 1999, but with the rather more prosaic name of Lavender Cottage. Sadly, however, it had been scheduled for imminent demolition to make way for industrial buildings.

Mac Cooper described his father in the draft of his intended autobiography.

He was a big-framed man, well over six feet and weighing about 14 stone, but devoid of surplus flesh. As a young man he had represented Hawkes Bay at rugby and it is one of my precious memories that he took me in 1921 to McLean Park in Napier to see the New Zealand Maori team narrowly beaten by the touring Springboks. Apart from his qualities as a farmer he was prominent in community affairs. He was a foundation member of the Havelock Town Board until his death and he was also a long serving member of the Napier Harbour Board. He was an active freemason and had his turn as a worshipful master of his lodge. Getting

'The Cosy Ship' – Mac Cooper's birthplace.
Re-named Lavender Cottage, this photograph was taken in 1999, shortly before its demolition.

into his gladrags for lodge meetings, described by him as 'going to ride the goat', was always an occasion.

These extra-mural activities are also referred to in Grant's book on Havelock North. On one notable occasion in 1912, the village organised a great Shakespearean pageant, the purpose of which was to raise money to liquidate the loan on the village hall. The *Forerunner*, a local monthly publication, recorded the two days events in some detail. Amongst those listed in the grand parade were Laurence Cooper dressed as Othello and brother Jack, who with Eric Phillips ran the coconut shies. The report goes on to say that the event was voted a huge success and long talked over by the happy participants until after the Second World War. The same book also records the names of the thirty-eight men who were killed in the 1914-1918 war, 'many of whom were sons of the early settlers of Havelock, reared and educated in the village before they were called upon to fight for their country'. Jack Cooper was one of those thirty-eight.

The name of L.T. Cooper is indeed listed as a member of the Town Board from 1912 to 1924, and perhaps more significantly, the Appendix of street names lists Cooper Street, with the citation 'for L.T. Cooper, who owned property in this area. He was a member of the first Havelock Domain Board appointed in 1906 and later became a member of the Town Board.'

In the garden of the 'Cosy Ship' with Mac and elder brothers Dick and Laurie aboard 'Nugget' – Tom holding the reins. 1912.

It therefore seems clear that Larry Cooper was a man of many parts and by the time his youngest son appeared on the scene his endeavours had begun to bear fruit. He had already purchased a seventy-acre block of good fertile land out of town on the road to Mangateretere. There was no house or buildings to speak of which meant living off site. Gradually the cropping gave way to the setting up of a dairy herd, followed by the direct retailing of milk into nearby Havelock North. The need to be on the farm with the stock meant that some of the family had to live in converted sheds with the very minimum of domestic facilities. The incentive therefore to build a permanent and spacious farmhouse rapidly developed into a major priority. With the help of a local building contractor, a fellow mason, Larry's family and anyone else who appeared on the scene, the project moved steadily from dream to reality and the family moved into Kianga Pai, Maori for 'Good Home', shortly after the end of the first world war.

Gordon McLaughlan in his splendidly illustrated history *The Farming of New Zealand*, describes farming at this time:

> As an uncompromising round of work for the majority and especially for those on small properties, who often only had a precarious financial grip. It was a life fraught with problems and uncertainties. The problems were usually practical ones and they produced many a brilliant do-it-yourselfer. The uncertainties were harrowing. The farmer, head of a nuclear family that had an accident or illness was in the hands of his friends and if the lore of the county seldom let him down incapacity was a constant worry.
>
> The family could suffer the educational and social disadvantages of remoteness, even when the roads were straightened, widened and sealed. There was no forty-hour week and many were long deprived of the comforts of electricity.

This picture, whilst true in some ways, was mitigated somewhat in the case of the Coopers in that they were within easy reach of a thriving community. They were also numerous enough to each contribute in their own way. The elder sons could help on the farm before and after school and the girls were fully involved with the domestic chores.

Another critically important factor was Sarah Ann's determination that their children should have the best education possible, even if it meant a delay in their financial contribution to the family. This may well have arisen from her own experience as a girl in the South Island and the apparent distaste her parents had had towards her own education. Her own father's view had been — girls did not go out to work, therefore education was a waste of time.

'Kianga Pai' – the Cooper farmhouse built shortly after the end of the First World War

'Kianga Pai' in 1999

Historian D.O.W. Hall wrote of the time:

The oddest fact about the vigorous first wave of settlers with their keen sense of the beauty and potentialities of the adopted country was their neglect of education. Men who had been well educated themselves, who indeed owed to their education and nurture the flexibility of mind which had given them their success as colonists, made inadequate provision for the education of their own children – though a few contracted out by simply exporting their children back to Britain for education there. One result of this neglect was a distinct grossness and lack of refinement of manners in the colonial-born generations, another their sense of conscious inferiority in contemplating the parent civilization of Britain, promoting a sense of abject, uncritical devotion.

Sarah Ann would have had none of that!

So with these forebears, what traits might reasonably have been expected in the genes of the young Mac Cooper? First and foremost, a sense of adventure and courage manifested in Grandfather James in particular, but also possibly a generic component within most antipodeans.

Second, a strength of character and determination to see things through. Both his parents appear to have possessed these assets in abundance. Third, an enthusiasm to become involved in the affairs of men, whether at work or at play, rather than stand by and let others make the decisions. And finally, a deep integrity and genuine sense of love for the family unit.

2

'. . . AND THEN THE SCHOOLBOY'

Mac Cooper's first chapter of his intended autobiography described his early school days and was entitled, again in Shakespearean mode: 'And then the schoolboy . . .'

According to sister Thelma, he was a hyperactive child and on the advice of the local doctor, their mother agreed that it would be sensible to send him to school when he was six, a year earlier than the then compulsory starting age. He wrote:

When I first crossed the portals of Mangateretere School, sometime in 1916, it was a modest two-roomed timber structure with an enclosed porch running the length of the building. There were rather primitive toilets and an open fronted shelter in the playground that gave protection during school breaks when the weather was bad. This was used only very occasionally for that part of Hawkes Bay enjoys a very good climate, particularly when assessed from the point of view of small boys. There were approximately seventy pupils varying in age from about five to fourteen drawn from within a radius of about three miles. Most walked to school but some had bicycles and a few had ponies, held by day in a small paddock which was part of the school grounds. These included a garden, the pride and joy of the headmistress, a redoubtable Scot by the name of Miss Balfour, who was better known to the two generations of pupils as Biddy Balfour. One of the privileges enjoyed by the better behaved was being dispatched to move the garden sprinkler during the hot dry spells that are so frequently experienced in Hawkes Bay. They also had strict instructions that the ground should be given a good soaking, for inadequate watering was worse than none at all.

About one third of the pupils were Maori who came from two Pas, Waipatu and Kohupatiki, that were in the catchment area. They were rather a moving population in that someone who appeared to be an established member of the school would suddenly disappear only to re-appear months later, explaining that he had been stopping with an aunt or uncle. They were bi-lingual, speaking their own melodious language to each other and English where it was necessary, sometimes with difficulty when it came to grammar.

We were a barefoot lot at school, Maori and Pakeha (Maori for white man) alike, and this was not due to poverty but choice, particularly in the summer. It was only when we boys were rounded up to go to church or if we went into town that we wore boots. Shoes were out completely because they were worn by either girls or sissies. The barefoot habit is a persistent one in New Zealand for when our daughters went to school they wore sandals but invariably one earned her mother's reproof through arriving back barefooted with one or more sandal dropped somewhere on the way home. Our schooldays started at 9 am and finished at 3.45 pm with an hour break at noon for lunch which usually consisted of sandwiches possibly supplemented with an apple, a slice of home made cake or a biscuit. There were no school meals as such, for dinner ladies had not yet been invented and we had to make do with whatever our mums were able to provide. As we sat round in our groups eating our lunches there was always interest in each other's rations, and envy too, if someone produced a chocolate biscuit, a banana or an orange, for those were luxuries indeed. There was a lot of sharing of lunches by the Maoris and often a child would be completely dependent on another's bounty. There was no demurring for share and share alike is basic to the Maori way of life, often to the detriment of economic progress in adult life.

Maori lunches really came into their own following a *tangi* which is the send off given by Maoris to their dead. Basically it consists of a period of organised wailing with women taking the leading role before the body is committed to the grave. Then comes a period of feasting with whole pigs and sheep, slabs of beef and vegetables cooked in the traditional manner of water on heated stones to produce steam in earth ovens. The length and intensity of the wailing and the size of the feast were related to the importance of the deceased.

Maoris have colossal appetites but, judging by the school lunches that followed *tangis*, there always seemed to be some left-overs which became an unholy mess of slices of pork, trifle and jellies in containers; but it all went down the same way.

True to Scottish educational tradition, Miss Balfour was a great believer in the three Rs, especially the use of English, and I shall always be grateful for the grounding she gave. Spelling exercises became something of a game and so did explaining the meaning of words and giving antonyms or synonyms or opposite genders of given words. She was also a great believer in mental arithmetic, no doubt a typically Scottish safeguard against

11

being taken in by shop assistants whose mental arithmetic tended to be in their favour. Biddy had a wonderful knack of developing even a modest aptitude for handwriting, drawing and painting. Maoris in particular seemed to excel but possibly the best in my time was a dull plodder rejoicing in the name of Ernest, who got sparing praise for his painting but hell for his other efforts. I had a fellow feeling but in reverse because I got hell for my painting and modest praise where he had failed.

Playtime activities were varied but the lack of numbers and the absence of teachers who could coach beginners limited the scope for team games. There were never enough of us of comparable size to form a rugby fifteen; nevertheless we played a form of seven a side rugby between ourselves during the winter months. It would not have been New Zealand otherwise for then as now rugby was a national religion.

I did not return to Mangateretere at the end of the long vacation in 1919. By a variety of specious arguments, I convinced my parents that it would be much better if I transferred to Havelock School. I believe there still exists somewhere in the family a crawling letter written to my mother who was then in hospital recovering from a goitre operation.

Thelma confirms this event. Mac was desperate to get involved in team games and could see no future staying at his first school. Miss Balfour on the other hand was reluctant to see her star pupil leave her charge. Eventually persistence paid off and his mother agreed. It also happened to come at a convenient time as Thelma herself had just completed her secondary schooling. She had been appointed as a probationary teacher at Havelock, a post which she was to hold for two years before going to the Teacher Training College in Wellington, where her elder sister Madge had started her nursing training.

The plan was for Mac to accompany Thelma to school in the pony drawn gig each day – a condition he readily agreed to as it opened up exciting new horizons on the sports field. He explained enthusiastically: 'It meant that I could enjoy organised football and cricket and play in competitions against the big schools in neighbouring Hastings, then a town of ten thousand souls and a vast metropolis in the eyes of a simple country boy.' Mac seemed to thrive at Havelock School and he continues his reminiscences thus:

Without in any way suggesting that Miss Balfour's teaching was inadequate this change had a good effect on my academic progress. This arose from having competition that was lacking in the smaller school. My main rival for first place in my final year was a boy who played no sport and in my philistine eyes he was something of a poofter! Apparently he ended up as a Presbyterian parson, but I am not holding that against him. I was second in the race but not to my adversary, for we were both pipped at the post by a quiet unobtrusive girl who was something of a dark horse. I did not begrudge her this triumph because for me, the really important issue was beating the other boy.

This mention of ecclesiastical destiny raises the question of my own religious education. We were all brought up to say our prayers on bended knee before we went to bed and the elder members of the family were required to attend Sunday school and also learn the shorter catechism. Here Mother set a good example to her offspring by also learning and reciting the catechism to the Minister's satisfaction and receiving a certificate to commemorate this achievement. She was a devout Presbyterian and while our home was still in Havelock the family were regular worshippers in the grey wooden church with its tall steeple, which sadly is no more because a few years ago it was demolished and replaced by an uninspiring up-market example of ecclestiastical architecture.

Sunday observance by the family changed markedly when we moved to Mangateretere as instead of a short walk to church it was necessary to load a trap and later a model T Ford with young Coopers and haul them off to their devotions. They, in turn, would endeavour to take avoiding action by disappearing into the blue immediately after breakfast on a Sunday morning returning in time for dinner, which was the really important meal of the week. Every now and then Mother would clamp down on our truancy. 'You are going to church with no nonsense today,' she would exclaim at family breakfast and we duly went, though this chap 'No Nonsense' became something of a family joke. Elements of our church intrigued me, in particular what happened to the collection money and the nature of the decorations sported by women on their hats. Artificial fruits looked so real that I felt they must be edible and with a small boy's regard for his stomach, I wondered why they had survived so long.

The bread and wine that was handed out at Communion was another challenge and on one occasion I snatched a bit of bread as it went by and had to be restrained from helping myself to the rationed thimble of wine. That didn't worry me as Mother always screwed up her face when she took her portion and I had the impression that it must have tasted like medicine. She was opposed to alcohol in any form and perhaps she did this in order to convey the impression that there was no pleasure to be derived from such imbibing, but alas, any effect on us was not very long lasting.

Though I could never be described either as a diligent or a devout churchgoer, those early Presbyterian services have left their mark. I have in consequence never obtained any real spiritual satisfaction from formal Church of England liturgies with their mechanical responses, which, I suspect, have no meaning to the participants in nine cases out of ten. When occasionally I attended a Church of Scotland service I got comfort from its simplicity and directness, which, for me, create the right atmosphere for an appraisal of life, and its relationships with other people, particularly those who are close to me. I am ashamed to say that during a C of E service I am more likely to be wondering how much longer it is going to last, particularly if the parson has an affected speaking manner.

Soon after I started school at Havelock we moved into our new home which totally altered our lifestyle. There were five bedrooms, a large living room which doubled up as a dining room on special occasions and a large kitchen with a pantry and the inevitable Shacklock wood burning stove, where most of our meals were taken. Above all there was a bathroom with hot and cold running water that also served the kitchen sink. Toilet facilities were still primitive but the closet was not isolated for it was discreetly incorporated in a building that was also a garage for the Model T Ford, which Dad acquired at about this time.

It must have seemed like paradise for Mother to return to a proper house after two years of camping on the farm, but this does not mean that life was easy for her, as there were still the family of nine to be fed and clothed. There was also preserving and conserving of fruits in season and the curing of three or more carcasses for bacon on top of other housework. In addition there was an ambitious garden which she and Dad had created, a proper wash house with a built in copper for boiling clothes plus a pair of wooden tubs for soaking and rinsing with a ringer fitted between them. This chore of weekly wash became Madge's responsibility while she was still at home and it was always her great ambition to be the first in the neighbourhood on a Monday morning to have a line full of washing. It was my job in those days to ensure there were sticks for the copper and also kindling for the kitchen stove and later the bigger wood that kept the fire going. I learned to use an axe early on in life but in those days everyone had to pull his weight as soon as he was old enough to make a useful contribution so there was no feeling of hardship. In any case we had many good things in our life by way of compensation such as watermelons, and a wide variety of tree-ripe fruits from the small orchard attached to the farm.

There was no electricity when we moved into Kianga Pai and we relied on candles and oil lamps though not without some envy of friends living in Napier and Hastings who had power. Electricity arrived in 1927 when the big hydroelectric scheme at Lake Waikaremoana (sea of rippling water) started to function.

The lines had been installed and the house had been wired six months previously in readiness for the great day. I will never forget the first night the power came on. It was about 5.30 pm on a winter's evening, when on my way back from school, I came round a bend on to the straight stretch of road leading to our house which seemed to be ablaze with light to the point that it almost seemed to be on fire. Brother Tom was so excited that he had switched on every light both inside and outside the house.

For all of us country dwellers, after three years of making do with lamps and candles, it was much more than a nine-day wonder. Incidentally New Zealand was far in advance of Britain in respect of rural electrification and in the late twenties less than five per cent of New Zealand homes lacked electrical power. This was remarkable, taking into account the low population density of a dominantly rural country and the cost of bringing power to isolated farms. Ironically we had electric light in the cowshed well before we had it in the house. A retail milk round not only necessitated an early morning start in the summer but in the winter as well when darkness was a handicap. Our parson at the time was possibly more interested in electrical manuals than he was in the bible. Not only did he suggest harnessing a dynamo to the Hornsby engine which powered the milking machine, but also he set about installing the system to make the Cooper cowshed an oasis of light in the small hours of a winter morning.

But all was not well in this hive of activity and it became increasingly clear that Mac's father Larry was becoming over tired, perhaps owing to the undoubted pressures of establishing the farm as well as the retail milk round. A heart weakness was diagnosed, and he was told that he must reduce his workload. As a consequence the retail business was sold and the family took on increased responsibility in an attempt to persuade Larry to slow down. In fact Mac took a part time job with the brothers who bought the milk round shortly after he moved to Havelock school, just before his twelfth birthday.

I was asked to help out temporarily but as the illness of one of the brothers turned out to be serious, I carried on for the next eighteen months. This was not unwillingly because I had been envious of my schoolmates with paper rounds and a half guinea weekly seemed like great riches to me. It involved an early start because we had to be on the road by 6 am to complete the round by 8.45 in time for school. Fortunately at the end of the school day there was Thelma with her gig and pony to provide transport home. Of necessity there was an early bed routine to end my day but I had no regrets about the commitment. Apart from doing what elder brothers had done before me it was a nice feeling for a Scot at heart to be salting away at least a pound a month in my Post Office Savings Account.

On top of this I got to know Havelock, which was steadily growing into a select residential area, particularly for retired people. I got to know the mean and the generous among them, especially at Christmas. Also the good and bad payers, as well as the clean and the not so clean

15

housewives. The latter were the bane of our lives because they were also the ones to complain about milk going sour which was not surprising when the receptacles they left for their milk so often contained the remnants of the previous day's delivery. I have through these experiences retained a sympathy for the 'pinta man' that will last to the grave.

The threat hanging over the Cooper household at the end of 1923 was tragically realised two days before Christmas. Larry had gone down to the farm after breakfast to do a few odd jobs and when, by mid morning, he had failed to return, it was assumed that he must have gone to a neighbouring farm to help the men harvest rye grass seed. As the day wore on and someone checked that he had not gone to the other farm, there suddenly dawned real concern. Mac and brother Tom began a systematic search of the farm buildings and eventually found Larry's body slumped in a tank that contained water for cooling the Hornsby engine. The engine had required replenishing and in the course of this he had accidentally dropped the bucket into the tank. His efforts to retrieve the bucket must have triggered a heart attack and he had collapsed into the tank. The post mortem revealed that there had been no water in his lungs. The family was devastated and Mac wrote:

> The effect on all of us was profound but for Mother it was completely devastating and I don't think she ever got over her loss. Theirs had been a perfect marriage and to lose her mate just at a time when the family was growing up and they were in a position to enjoy life together after the hard slog of the early years was a truly cruel blow. The wound that it left never truly healed.
>
> Dad's familiar name Larry was on her lips when she smiled as she passed away seven years later as a consequence of an embolism following an operation. Madge was at her bedside.
>
> Dad died soon after my thirteenth birthday so I was not old enough to know him properly or to develop fully the sort of relationship one hopes to see between father and son. Even so he had already earned my respect for what he was and what he had done. His funeral was one of the largest that Havelock had seen and this was a reflection of the esteem in which he was held. There were many of all sorts of station in life who had cause to be grateful for the help he had given and their messages of sympathy were a great source of comfort to Mother.

Secondary schooling for Mac Cooper commenced in 1924 when he enrolled at Napier Boys High. This meant an even earlier start. He saved up to buy a bicycle, which replaced the faithful but impracticable pony and trap to take him to Whakatu station, some three miles away. The ten-mile train journey to Napier, followed by a fifteen-minute walk, meant an hour travelling at each end of the day. Arriving home usually after 5.30 pm, two hours of homework were theoretically sandwiched between an evening meal and bed. But in practice the homework tended to be much shorter unless English literature required the reading of some interesting book. He explains the consequences of such a selective preference:

> In common with most of my train fellows we used the half-hour of the morning train journey to make good deficiencies in our previous evening's preparation. But too often the shortfall caught me out and I had to face either detention with a late train home or a few strokes of the cane, which was a preferable punishment because it was over quickly. Our maths master, Mathew Alexander, better known as 'Rough Alec', was a great believer in the cane. 'I am sorry, Cooper,' he would say with his distinct Scottish accent, 'I'll give you four cuts.' They did my geometry no end of good.

As Mac progressed through the school, the inevitable matriculation examinations loomed at the end of his third year. He was not expected to excel by his then form master. However, more to prove him wrong rather than with any other deep-seated motivation, all seven subjects were safely negotiated. This success led to promotion to the sixth form where there was an opportunity to widen the subject base with the aim of attaining a leaving certificate which carried with it a small university bursary. Although Latin as a subject was only required for those intending to study medicine, Mac added it to his list. It is also probable that this subject was known to be a requirement to qualify for a veterinary course.

Animals and their welfare had long been a subject of great interest to Mac Cooper. His father was well known in farming circles for his innate skill in this field, so much so that neighbouring farmers were for ever seeking his opinion as to how they should treat livestock ailments. No doubt the local vet was more expensive.

This influence obviously rubbed off on his youngest son and from an early age Mac had set his heart on a professional career as a veterinarian. Had not Larry died prematurely, this ambition might have been realistic, but clearly he could see that the financial demands of such a choice were out of the question as he neared the end of his time at school. It seems, overall, that he felt his academic progress fell short of his aspirations during the rest of his time at Napier High:

> In some respects the two years in the sixth form were not used to best advantage. It is true that on one side I became a member of the first XV, a school prefect, head of my house and in my last year, Company Sergeant Major. But it was on the academic side I failed to make the progress that

would have made my first year at university less demanding and it was not entirely my fault.

When I discovered that my first choice, a veterinary degree, was out of the question because the nearest training centre was at the University of Sydney, costing money well beyond even my wildest dreams, I opted to read agriculture, which had recently become a university discipline.

I was put on a not very challenging course of elementary agriculture because the school did not then appreciate that the right sixth form foundation for a professional training in agriculture is a good grounding in such basic sciences as chemistry, physics and biology. Fortunately I continued with inorganic chemistry and some aspects of physics that had not been taught previously, but there was no biology or organic chemistry offered at that time. Sensibly I continued to study English literature, where I was fortunate enough to have an outstanding teacher.

Sadly in this age of specialisation and pressure cooker preparation for university entrance examinations, there are too many sixth form students who have dropped English Language and Literature to concentrate on science subjects and the result shows up in their inability to use their native language properly. Some are almost illiterate when they put pen to paper.

My university ambitions undoubtedly had their genesis in the favourable outcome of my matriculation effort. Thelma had not long finished her teacher training in Wellington where she had contacts with the university and she appreciated the importance of higher education. Above all it was Mother who encouraged me in my dream of going to varsity. She was a typical Scot in her respect for education and the thought that her youngest might one day be a graduate obviously appealed to her. I think her secret hope was that I might become a minister of the Kirk, because of the high regard she had for several men of the cloth whom she had known, but I think she also realised that this was most unlikely as my interest in evolution was turning me into an agnostic of sorts.

Certainly my unconventional views at that time caused great alarm in the mind of a neighbouring farmer; a middle-aged bachelor dubbed 'Gloom' by the family, because that was his disposition. Nothing was ever right in his world until he saw the light and became an unquestioning student of the scriptures, particularly the Old Testament. With a callow youth's characteristic lack of respect for an older man's treasured beliefs, I refuted the teaching of Genesis and expounded my own half-baked ideas. Gloom's reaction to my blasphemy was on a par with that of Captain Footsore of *Beagle* fame when Darwin advanced his theories on the evolution of the species. Mother received a report from brother Tom, who was rather impressionable in matters like this, that young Mac was rapidly going to the devil and something should be done about it before it was too late.

Gloom was not the only neighbour concerned about my future. It was a source of great wonderment to many in our locality that anyone should remain at school beyond the age of fifteen and here was I, the third of our family after Thelma and Dick, to go on till after our eighteenth birthdays.

Mac Cooper finally left school in December 1928 at the age of eighteen determined to enter Massey Agricultural College, which had just recently opened. But in order to qualify for entry to the degree course, he had to embark on an intermediary year at Victoria College in Wellington, this being a constituent body within the University of New Zealand. Here the subjects were the basic sciences, at which he had to work hard, to make up both for the shortcomings in his previous school syllabus and a lack of dedication to the task in his final year.

Life in this city of one hundred thousand people and the challenge of living away from home in the country constituted a significant change in lifestyle. The responsibility was squarely on the student's shoulders, both to find suitable digs and to keep up with the work. Mac also began to appreciate the considerable sacrifices that were being made by the family on his behalf, as the only grant was the £20 per year which he had been awarded following his obtaining his School Higher Leaving Certificate. This was insufficient to pay for his tuition fees, let alone keeping him fed and watered. So the family had to help him make ends meet.

Most of the 1800 students in the College were there on a part time basis. Apart from the aforementioned grant, there was no other financial help available to cover costs, save that provided by parents, only a small proportion of whom were in a position to provide significant help. Full time attendance was obligatory for those who were to study medicine or engineering and indeed this was the case for agriculture. In this particular year there were only five students enrolled for a subsequent course in agriculture, one of whom was destined for Lincoln in the South Island and the remainder for Massey.

Life was not all hard work and the prospect of playing rugby in one of the senior university sides was a huge incentive. With the track record of two years in the school 1st XV under his belt and many games with his brothers in the local rugby team back home (within which, when tired after a long series of play, they were apt to 'recooperate'), Mac was exceedingly disappointed to find he was selected towards the bottom of the eight sides which were fielded on Saturday afternoons.

Persevering, however, he ended up in the thirds and once represented the second team. He was bitter about that experience, because the policy at the time allowed old students to play for the university long after they had left and taken up careers, so preventing many up and coming youngsters the opportunity to play first class rugby for the first time. Indeed when he returned to Wellington some years later, he opted to play for the club side rather than the university which still persisted with this policy. With a markedly enhanced reputation by then, he was roundly criticised in some quarters for being disloyal to the university cause. But he had no regrets and he made sure that everyone knew why he made this choice in the belief that if others followed his example, then more opportunities would come the way of aspiring young hopefuls at the start of their rugby careers.

Another activity which Mac Cooper enjoyed was military service. No doubt this first arose back on the farm when, as a boy, he played trench warfare with his elder brothers. At school he was a member of the school cadet force and rose to the heights of Company Sergeant Major. By the time he got to Wellington, military training was still compulsory for all young able-bodied men. As a second lieutenant he had charge of a motley collection of fresh faced youths to lick into shape during their fortnightly sessions at a training camp on the outskirts of Wellington.

He confessed to feeling 'a bit of a nana' dressed up in bulging breeches, putties, a jacket with shining brass buttons, a carefully polished Sam Brown and a Boy Scout type of hat which the New Zealand forces then favoured. Also, a swagger cane that had been presented at school in recognition of achieving commissioned rank.

Victoria College was not without its social life, some mixed, some distinctly for men only. The Hongi Club took a leading role in organising the annual capping procession through the main streets of Wellington with students appropriately dressed or undressed as the situation demanded for some take-off of an institution or event that had recently been in the news. Mac's modest contribution in 1929 was that of a drummer in the procession band. For this he wore a bowler hat, as a fellow student thought he might look a little odd in it. Mac asserts: 'He was dead right. Ever since, for funerals or British Legion Remembrance Day parades, I stand naked-headed and ashamed beside my be-hatted companions who have the right kind of physiognomy for wearing bowler hats.'

The Hongi Club also ran smoke concerts, which were on the bawdy side, perhaps a shock to the sensibilities of the young Cooper. He opines, somewhat pompously, that the leading lights of the Club 'set a rather vulgar tone at these events'. A far cry from Mac's later reputed mastery of all eighty verses of 'Eskimo Nell'!

Maybe he was a clean living lad in those days enjoying only a couple of beers after a rugby match. One vice, however, he owned up to was smoking. Not heavy at first, but as he himself admitted, the first signs of becoming an addict were there. Owing to his impecunious circumstances, he tended to roll his own. This was a technique which involved purchasing the tobacco and the packet of papers separately. In expert hands the resulting product passed as a reasonable impersonation of a cigarette. Mac's efforts were less than expert with the result that either it wouldn't light owing to over generous licking of the two ends of the paper or it set off like a roman candle scattering burning fragments onto the carpet.

As this intermediary year drew to its close, the examinations remained as the final hurdle. Although physics was fairly borderline, the results were successful, enabling another step on the road towards a degree course at Massey.

At the end of the final term Mac returned to Hastings and took a job at the Heretaunga Cooperative Dairy Butter factory. His father had previously been a director at the factory for many years and this helped in getting the job, no easy task as the world recession took hold, making employment extremely difficult to come by. He had opted for the Dairy Science course at Massey in the mistaken belief that this would involve the practical side of dairy farm management, which was his first love. It transpired however that this was designed for those who wished to go into the milk processing side of the industry and hence the work at the butter factory. Whilst welcome from an income point of view, it was unlikely to provide much useful experience for his chosen subject.

Early in 1930, Mac's mother was admitted to hospital for an operation to remove her spleen. Fortunately the hospital in Hastings was close to the factory where Mac was working which enabled him to visit her regularly. All seemed to be proceeding according to plan when suddenly she suffered a relapse attributed to an embolism. She died on 9 January, aged fifty-nine. For the second time within seven years the family were grief stricken by this totally unexpected bereavement. Mac had become very close to his mother since his father died and he wrote with warmth and affection about their relationship:

After the initial shock I came gradually to realise that the loss was one of flesh not spirit, for hers will remain with me always. It was her faith in my ability that took me to university and at least I gave her the pleasure of knowing that I had succeeded in passing the first set of papers because the results were announced shortly before she went into hospital.

I can only imagine the pride that she would have had in my subsequent achievements and yet somehow I feel that she was aware of all these things and much more besides. It is my hope that those who follow after me in the family will have some inkling of the woman I am so proud to call my mother.

Mac left the butter factory after a spell of eighteen weeks and was presented with a pair of London made pipes as a leaving present. There were only about fifteen employees and with the hard times which all were experiencing, it was a generous gesture to a man for whom they had a high regard.

After his mother's death the farm was sold and Mac's brothers went their separate ways. Jim went sheep farming, Tom got a job managing for someone else and Laurie and Dick bought a truck and started a contracting business. The 'Cosy Ship', however, still belonged to the family until the mid 1930s, with the girls running it for any of their brothers who needed a home.

Unfortunately Mac Cooper's draft autobiography ends at his point of departure for Massey, where prior to the commencement of his degree course, he had to complete a full year of farm practical work. Professor Riddet had offered him the job of pigman for this year, provided he successfully completed a four-month apprenticeship under a certain Campbell McMeekan.

3

MASSEY

Massey Agricultural College was opened officially on 20 March 1928, almost fifty years after its counterpart, Lincoln College in the South Island. The author of *Massey – Its Early Years*, T.W.H. Brooking, describes the sequence of events which resulted in its remarkable gestation period.

At the turn of the century it was becoming increasingly obvious that New Zealand was falling behind other developed countries with regard to the application of scientific advances into practical farming systems. As a nation increasingly dependent on agricultural exports, the need to boost productivity and improve technical expertise was undeniable. How best to provide the necessary human resources with regard to agricultural education and research, however, was a subject about which political and academic argument had raged for years. Confusion reigned over whether the emphasis should be on the practical farm school as opposed to the university department. Just as emotional was the issue of location. Should the additional resource be the setting up of a North Island College or the expansion of Lincoln in the South Island? If it were in the North Island, should it be a new establishment or an addition to Victoria College in Wellington or the University of Auckland?

It was not until 1912 when William Massey's Reform Ministry came to power that there was a firm commitment to put the issue at the top of the Government agenda. Massey was an advocate of a college in the North Island working in conjunction with the other universities and offering degree courses in agriculture.

Political intrigue and indecision continued until after the First World War and it was not until 1924 when Geoffrey Peren was appointed to the Walter Clark Buchanan Chair of Agriculture at Victoria University College that real progress began to be made. Peren was confronted with a department consisting of a small room in the Physics building, barely large enough to hold six students. He had no other facilities and in his own words Peren remarked, 'His must be one of the most extraordinary Schools of Agriculture ever to accept students.' The Government were unyielding to his requests for further funding and to make matters even more confusing, Auckland University proceeded to create a new chair in agriculture.

Peren decided, however, to take matters into his own hands and launched a public relations campaign to drum up support for a single establishment, maintaining that the North Island could only afford one department and having enrolled twelve students, it should be at Victoria College. Luckily for Peren, a Royal Commission in 1925 supported his views and, encouraged by this rather unexpected turn of events, he proceeded to roundly criticise the Board of Agriculture for 'dragging its feet'.

The next significant move came with the appointment of Professor William Riddet from the West of Scotland College of Agriculture to the Chair at Auckland. He was a specialist in dairying and obviously determined to develop the Faculty in Auckland with Ruakura as the associated site for farm and research activity. Peren, anxious to encourage amalgamation, proposed a joint conference between the two establishments to consider the setting up of one school of agriculture at an alternative venue. Riddet meanwhile had had time to evaluate the options and realised the danger of the Government playing one school off against the other in financial terms; also he had severe reservations about the distances involved between his department in Auckland and Ruakura in Hamilton.

The critical change, which enabled progress then to accelerate, was the working together of Peren and Riddet. They wanted things to happen and as so often is the case, people with vision and determination were able to put forward reasoned argument in an attempt to narrow the ongoing political differences. They worked jointly on the content of a four-year degree course to Bachelor of Agricultural Science. Following another joint conference at Wellington, it was resolved that one university school of agriculture would be established in the Palmerston North area, because of its good communications and central location to the whole of New Zealand. It would concentrate on animal husbandry and leave Lincoln, with its drier climate, to provide for the needs of farmers in the South Island who were engaged in arable and root farming.

Clearly this specification meant a college farm suited to dairy and sheep production and eventually a deal was agreed for the purchase of the Batchelar property extending to some 820 acres near Palmerston North. After negotiation the price was fixed at £60 per acre. But the political battle against the new initiative was by no means over. The South Island MPs staged an orchestrated revolt against the New Zealand College of Agriculture Bill in 1926 asserting that the public expenditure was unwarranted as there would not be the jobs available for those who graduated. They also accused the proponents of the Bill that their only interest was parochial jealousy against Lincoln College. In the event the Bill was eventually passed owing to the majority of the Labour party, but any hope of unanimity of will seemed as far away as ever.

Meanwhile Professors Peren and Riddet were getting to grips with the installation of the farm infrastructure such as fencing, roads, tree planting and water supplies. Peren was appointed acting Principal and advertisements were placed for staff in the UK, Canada, America and Australia. Salaries offered were higher than those prevailing in the countries concerned, causing some criticism but also ensuring an excellent response from highly qualified and capable applicants.

In 1927 two more Bills were introduced to try and assuage continuing political unrest. The first was designed to appease the South Island opposition and it proposed equal university status for Lincoln College as well as increasing grants for upgrading their research facilities. The second was a masterstroke in that it altered the name from the originally proposed New Zealand Agricultural College to Massey Agricultural College, as well as granting university status. There is nothing quite so emotive as the choice or change of names for organisations and a name which was associated directly with a universally revered farmer statesman, himself an advocate of the College, helped considerably to cool the temperature.

With opening day approaching in 1928, there was much building activity in

Massey Agricultural College and grounds in the 1930s. Swimming pool under construction, bottom left.
(Massey Archives)

25

progress. The old McHardy homestead, which with thirty acres had been added to the farm purchase once the main deal had gone through, was physically split in two. This was achieved by means of an ingenious sequence of rollers; one part was wheeled away to become the Principal's residence and the other became a temporary teaching building. The farms were stocked with a nucleus dairy herd and sheep flock, adequate farm machinery had been purchased and course contents finalised for degree and diploma students. Both professors were released from their responsibilities at Auckland and Victoria College and all systems were set to go as the first eighty-four students enrolled.

O.J. Hawkin, the Minister of Agriculture, invited to do the honours at the official opening, referred to 'the long and strenuous fight to establish the College'. To many this must have ranked as one of the understatements of the century. But perhaps the Mayor of Palmerston North was nearest the mark when he predicted that the site would eventually contain a 'majestic pile of university buildings'.

The thirties were a difficult time not only for the fledgling college, but for everyone in New Zealand as the depression, triggered by the collapse of the American economy, reverberated around the world. John Mulgan, a New Zealander and later a contemporary of Mac Cooper's at Oxford, wrote in his book *Report on Experience*:

Massey University campus in the 1990s

26

In 1931 everyone was still talking about the depression as if it were a rainstorm that would blow over and leave the earth green and smiling and fresher than it was before. Towards the end of 1931 the bottom dropped out of the markets that supported us. After that people spoke about the depression as something rather more than a rainstorm, as a national calamity that had begun to affect their lives. People resented the depression, not only because it did them harm but also because they couldn't understand it. It settled on New Zealand like a new and unwanted stranger, a grey and ghastly visitor to the house.

New Zealand was a primary producing country. We sold wool and butter and frozen meat on a British market and prices fell with the British pound. Prices of primary products, the economists said, tend to fall first and fall farthest. It seems moreover, that they took longer to rise again.

We lived on the end of a pendulum swung by a trade cycle that no one had bothered to assess before and now that we came to recognise its existence, there didn't seem much possibility of controlling the trade cycle from our remote end of the pendulum.

At Massey the consequence was that Government funds were drastically cut and staff salaries reduced by twenty per cent over two years. It was even decided to discontinue fire insurance to save the premium! Reducing staff numbers made some economies, but the main thrust of survival lay in the innovation and commitment which everyone undertook to ensure that life went on. Self-help became an ethos which served the College well over many years. It welded both staff and students together so that buildings continued to go up, the farms began to contribute returns and the students learnt fast through practical involvement in a vast range of different activities. As a consequence, the reputation of the establishment began to increase as manufacturers of commercial products made donations in kind to enhance their own publicity. Local firms gave books to the library; Breed Societies donated stud stock and prizes for students and a number of private beneficiaries left money to the College.

This then was the background to Massey Agricultural College when Mac Cooper arrived to do his practical year as an assistant pigman under McMeekan's supervision. Recalling these early days in a letter to Gordon McLauchlan, McMeekan's biographer, Mac Cooper writes:

We started the summer together living in the pigman's cottage fending for ourselves. There I quickly found that Mac had yet another attribute – he was a first class cook and we lived well on the weekly budget of 30 shillings for food and electricity. Admittedly there were perks: cracked eggs at a nominal price from the poultry section, milk for 2*d.* per pint, and pork fillets which were scrounged when we went to Longburn to measure experimental pigs. Mac was nothing if not a good scrounger and I learned a lot from him on this score.

During that hot summer, on a very hot Christmas Eve, Mac went to a farm beyond Longburn to collect a Large White sow from a farmer who had imported breeding stock from Canada. It was the first Canadian Large White Massey owned and Mac set out with Dolly, a stylish black mare, and the piggery cart, conscious of the importance of the occasion.

He arrived back late and completely dispirited. Apparently half way home the sow worked her way loose through the restraining net over the cart. He had had the unenviable task of leading the horse and driving the sow, which was as hot and bothered as Mac and tried to take refuge under every bush on the way. Eventually he got her to the Manawatu Bridge, which was then an old wooden structure. To Mac's dismay he held up a lot of jeering Massey staff returning from work as his procession slowly passed over the bridge. Eventually with some help he got her onto the roadside cutting and rolled her back into the cart. A few minutes later he was safely on his way once more. Professor Riddet passed him, never suspecting that his precious sow, as well as his student pigman, had been reduced to a state of physical and mental exhaustion.

Mac had his good qualities as a housemate but he also had his drawbacks. He worked on the principle of the first dressed being the best dressed and many a time I came home from work with a social engagement that night and Mac was away before me wearing my one and only clean shirt. The recompense was that he would often feed the pigs if I was late back from my rugby.

In his unpublished autobiography, Cooper refers to an incident which occurred while he was working with McMeekan in the piggery.

We were weighing pigs on the College farm as part of an experiment, when McMeekan accused me of deliberately rocking the platform scales. Suddenly we both realised that a severe 'quake' was in progress. Little did I think that the epicentre was in my own home area of Hawkes Bay over a hundred miles away. In fact the earthquake measured 7.9 on the Richter scale, claiming 258 lives and devastating vast areas in Napier, Hastings and surrounding neighbourhoods. It was described as New Zealand's worst natural disaster, destroying some $7.5 million dollars worth of property and infrastructure.

Fortunately there were no casualties in the area of Havelock North and none of Mac's family was hurt. But the pupils and staff at his old school in Napier had a very lucky escape. As it was the first day of term, virtually the whole school were outside on the playing fields engaged in military training whilst a handful of staff completed administrative tasks in preparation for the academic term. An hour earlier everyone would have been in the recently built Assembly Hall, which had a heavy reinforced concrete roof. It collapsed in one piece flattening everything inside when the earthquake struck shortly before 11am.

McMeekan, also better known to everyone as Mac, had been one of the foundation students at Massey when it officially opened and he had been the natural leader spearheading the drive to help create a university rather than a farm college atmosphere. There was no tradition, nor was there any self-confidence within the establishment about its new identity. McLauchlan paints the picture of the young McMeekan at Massey as superficially an 'extrovert clown, full of energy, the character, the hard case. They saw the manic front of the man who needed to be noticed, to be liked. He was always up or down, fast or dead slow. He was furiously engaged as often as possible in interaction with people.' There is no doubt that McMeekan had an impressive intellect together with a compelling, if not at times overpowering personality. He was quick-witted, sharp tongued and outspoken as well as having something of an irreverent attitude to people in authority. He was at the forefront of much high jinks and student pranks and seemed to get away with them, which greatly impressed his student contemporaries.

One of McMeekan's lifelong friends, Bill Hamilton, recalls: 'He said what

Massey Agricultural College Rugby XV, 1930.
Mac Cooper third from left, middle row.

he meant and this didn't endear him to politicians at large. I don't think he made enemies with his outspokenness. Enemies aren't the right word, but he was treated warily, like a dog. You never knew who he was going to bite next.'

McMeekan involved himself in all aspects of student life. He was elected as the first President of the newly formed Students' Association and as there was no sporting tradition or social club activity, or any establishment to confront, he had a free rein. He was instrumental in getting the first rugby side onto the field in 1929, having first procured a ball as a result of a sweepstake on the outcome of the test series then being played by the All Blacks in South Africa. As a ten stone, and in the words of his first wife, 'an emaciated' front row forward, he played his heart out. But it was self evident this was not his game and as soon as numbers enabled a wider choice of player, he happily joined in the games at a lower level. He did, however, remain involved as President of the club in later years and also represented the college interests in the Manawatu District Rugby Union.

Towards the end of his degree course, McMeekan was put in charge of the experimental pig farm and became an assistant lecturer after he graduated in 1932. At this point he took over responsibility of the pig husbandry department and also worked closely under Riddet's supervision with the dairy herd of 140 Friesian, Ayrshire and Jersey cows.

There can be no doubt that McMeekan made a huge impression on the young Cooper, almost to the point of hero worship. Here he was working alongside someone who was effectively the 'Head Boy', bright, popular and enthusiastic in every sphere of activity. Cooper had many of the same qualities and also the ambition to make an early mark on College life, but he had a different manner, a more quiet, thoughtful approach, perhaps rather in awe of his new mentor. McMeekan, for his part, recognised in Mac Cooper the leadership potential and the easy way he had with people. He could also lead by example. Here was someone whom McMeekan could groom to take over the mantle of responsibility as he neared the end of his tenure as President of the Students' Association. He might not have quite the same intellect but he was an infinitely better rugby player!

As a young man, Mac Cooper was a striking figure, tall, powerful and sharply featured. He was quick on the uptake, an enthusiast and a sportsman. Having lost both his parents he had a maturity and independence of spirit in advance of his years. He also had a little worldly experience having lived for a year as a student in Wellington and he had worked on the College farm for over twelve months. In short, an impartial observer would have been right, even at this stage, to single him out as having significant talent and potential.

That the relationship between the two Macs blossomed over the years was no surprise to anyone. Each in his own way made significant contributions to worldwide agriculture as their careers developed and diverged. McMeekan was more the 'ideas man', famous for resurrecting Ruakura to prominence in animal research. He evaluated problems quickly and imposed solutions, which if they involved trampling over hurdles or other people, he put down as the price of progress. Cooper, who remained in education, was more of a people person. He was an outstanding communicator and someone who had the rare gift of being able to fire students' imagination. He was a motivator and a stimulator. Some might argue that they were amongst the most influential agriculturists to emanate from New Zealand since the war. Whether true or not, to a generation of animal production men throughout the world, they were household names and left a lasting impression.

July 1931 saw the first publication of the Student's Association journal, known as *The Bleat*. The editor is listed as one M.M. Cooper.

A history of the Association, albeit short, describes the efforts made by the foundation students under McMeekan's leadership to draw up the constitution or 'McMeekan's bible' as it was known. It also records the very brief appearance of the rather grand Orchestral Society, which apparently faded into oblivion following its debut at the first College Ball. The Tramping Club, however, had stouter hearts and went from strength to strength, whereas the oddly named Karati Club existed to 'investigate the velocity at which beer can flow over mucous membranes'.

Massey colours were originally chosen as royal blue and gold, referencing the link and affiliation to both Auckland and Victoria University colleges. Some objections arose from Otago, however, in that this combination clashed with their colours. The Principal then decreed in the students' absence that saxe blue and wine red would be substituted. At least it shortened a long and tortuous debate. Another subject that took time to finalise was what now might be described as the College logo. The article recounts that a chance find in an old textbook depicting a noble ram copied from an ancient Egyptian fresco, ornamented with horns and a jaunty tuft of beard, filled the bill admirably.

The editorial in this first edition was what amounted to a 'call to arms' in rather flowery and evangelical style. Cooper's message was: 'We have the buildings and the farm, but we have no tradition upon which to shape our culture.' He wrote:

> The study of agriculture is a business and it requires a counter balance. The logical counter balance is an enthusiastic interest in the lighter side of college. An active Students' Association with a number of affiliated clubs has in its hands the social well being of the student.
>
> It now lies with the individual whether college is a succession of heavy lectures or a period of happy memories, hard work intermingled with those relaxations characteristic of an undergraduate. There is far more in college life than its academic associations.
>
> Every student, whether diploma or degree, has a common interest to

uphold the good name of the college. The justification of this college lies not in the examination results but in the character of the graduates it produces. Initiative and reliability are not the fruits of the lecture room. The men who have entered heart and soul into their student offices are far more likely to have these traits developed than the intensely studious who have confined themselves to their course.

He ends with the clarion call: 'We ask you to extend your team spirit to its utmost and acting as a whole raise Massey Agricultural College to the heights which our country demands.' This critical balance between hard work both on and off the field was a precursor to one of Mac Cooper's lasting hallmarks. He maintained that university education should focus on the rounded approach, an involvement in a wide range of activities and an opportunity for individuals to experience everything life has to offer.

By 1932 Cooper had taken over from McMeekan as President of the Association and he also remained as editor of *Bleat*. The list of clubs and activities was already expanding and as well as rugby, Mac had joined the tennis club and the Debating Society. In his editorial that year, he took a swing at the critics of the college, who, judging by the remarks he made, must have been pretty disillusioned about its early impact, results and accountability. In those depression years such a view is hardly surprising as hard-pressed farmers observed student activities from a distance. Rather than keep his head down, however, Cooper reveals his preference for stating what he regarded as a few home truths.

Why won't the farming community give the college a fair chance to justify its establishment? Persistently we are assailed by queries and statements from members of farming associations: 'What results has Massey College produced?' 'Why doesn't the College produce a balance sheet?' 'The biggest white elephant with which this country has been settled.' These are typical remarks and they do not indicate a great deal of sympathy towards, or co-operation with, the objects of the College on the part of our farmers.

Any demand for a balance sheet from the College is unreasonable. Neither education nor research, which are the main functions of the College, can be expressed in terms of pounds, shillings and pence. At what amount would the assessors value the collective knowledge of sixty students on such a cardinal point as top dressing? This institution is an extremely youthful one. Yet hardly has the paint dried on the front gates, and we have a general outcry for 'results'. To anyone in the slightest conversant with the difficulties of scientific research, it will be obvious that results cannot be published unless they have been substantiated by several years of repeated experiment. No real scientist can afford to jump to conclusions; it is equally absurd as a judge delivering a verdict before he has heard the case for the defence. Agriculture is so involved and is dependent on so many factors and complexities that it is impossible to solve its problems without years of patient work. Farming is not so simple as some farmers are inclined to think.

Cooper goes on to explain to his readers that it is education of the mind rather than basic research which is the primary aim of the College and this takes time. He returns to his evangelical theme of the previous year at the end of his article and urges the students to: 'Put your shoulders to the wheel and by stint of steady hard work, benefit yourself and increase the prestige of the College. Then you may come back forty years later and say to yourself, "I helped to make the name of Massey mean something more than a stately pile of bricks set amongst Nature's grandeur." God knows, it has been a hard enough row to hoe, with all these fence-sitters praying to their own particular devils that the College be damned.'

How widely read the publication was is not known; however it is interesting to see an Editorial Note, or in modern parlance, a 'disclaimer', which followed this article state:

The quality of the articles and their personal nature may not be regarded with favour by some readers of this publication. We are not attempting to make excuses for the quality of the material, but we desire to remind any critics that this is a purely agricultural college and the literary benefits of an Arts course are denied us. We have through this to resort to personalities to a greater extent than may be desirable, but at the same time we wish to reserve the student privilege of being critical. We should like to remind readers of this student right.

For a twenty-two-year-old it was controversial stuff and one wonders what the authorities thought about this frontal attack. Perhaps McMeekan remained a significant influence in the background. By this time he would have been on the staff as an Assistant Lecturer and may well have been looking for ways to wind up those on the outside that took such a negative stance. Whatever the internal politics, the evidence is there that even at this stage in Cooper's career, he had the courage to say what he believed, whether or not others engineered it with an axe to grind.

Despite the depression and the less than enthusiastic support from some of the local farming community, Massey continued to make headway. Improvements to the farm continued throughout the thirties with additional land being acquired and new teaching, research and student accommodation slowly materialised. The main building, still the focal point of the campus, was opened in 1931 by the then Governor General, Lord Bledisloe, himself a great supporter of scientific training within agriculture. Fortunately this building was completed before the depression put paid to further expansion plans and it meant that the teaching and laboratory facilities were now of an acceptable standard.

The College however remained small with an annual intake of under a hundred students and a staff of about twenty. It was not until the 1960s that Massey was able to change gear and move into a multi-faculty university. Now of course, it is a massive campus with some 30,000 students, most of whom are engaged in extra-mural courses covering the full range of educational opportunity. One of the most recent developments is the building of a National Institute of Rugby, perhaps an indication of how seriously and scientific the professional game has become in New Zealand. In addition to its playing and training areas, it will have residential facilities and direct links into academic courses such as nutrition and sports science.

Whilst the downsides caused by the depression were felt throughout the college, there was one benefit which the degree students enjoyed to their distinct advantage, this being class size and the ratio of teaching staff to students. Rarely were there more than ten students to a class so those lectures were more like seminars or tutorials, where individuals could play a full part in the subjects under discussion. Similarly with laboratory and practical work the effectiveness of staff input and participation in the process of learning was of very considerable benefit. Mac Cooper was a conscientious student and took the various annual examinations in his stride.

In his final year as editor of *Bleat*, he widens the perspective in his editorial to the global contradictions in agriculture exaggerated as a direct result of the all-pervading depression. There are almost the signs of a communist ideology creeping unwillingly into his philosophy, not an unusual trend to be found in student thinking in those days, when the prospect of gainful employment was distant indeed.

Obviously our economic system is obsolete. It has been crippled by trying to pay off capital that has been blown to pieces in France. But apart from this it has failed to adjust itself rapidly enough to the necessities of a quickly changing world. An entirely new balance between labour and capital is called for; a new method, which will convert the advances in science and invention into a blessing instead of a curse. Science has taken us up an Everest of accomplishment without giving us time to adjust ourselves to the rarefied air. We need a science of adjustment as well as a science of production.

It is not to be wondered at that we young men are worried about our futures. Every road seems to be a blind road while to many of us there seems to be no starting point at all to our journey through life. There is hardly a lucrative opening in agriculture and we as students cannot but be apprehensive of what the future holds. Those of us who can forget the dogmas of capitalism and adapt ourselves quickly to the new system are going to come out best. Capitalism is now in itself insufficient. It requires a very handsome admixture of Stalinism to make it adequate but the proportions of the two, time alone will determine.

I am not advocating communism. It is a distant world remote from this world of ours, possible only in a world of standardised souls, each with a common object, the community welfare and having an absolute selflessness. We are still, thank God, individual and we still have our own ambitions to achieve and our destinies to work out.

There has got to be wholesale rationalisation of production, distribution and exchange. Our world cannot afford overlapping. We all can't be professors of agriculture; there has to be someone to distribute the farmyard manure.

Under a headline 'The Editor has something to say', immediately following this editorial, Cooper bids farewell to his readership, thanking colleagues for the efforts they have put in while *Bleat* has been under his stewardship. Again he stresses that whilst some have criticised the publication for being over the top on occasions and disrespectful to those in authority, his aim has in no way been malicious. His colleagues and he have merely been exercising their youthful privilege of being able to make honest criticism. He urges older readers who may feel aggrieved to read it again with a pinch of salt. He ends by hoping that *Bleat* will continue to be published as he feels 'it is one of our most potent factors in the building up of tradition'.

One other light-hearted piece in the 1933 edition contains five cartoons of College personalities in a Rogues Gallery. Cooper, with angel wings sticking out in place of ears, has the 'Wanted by Scotland Yard' headline and the following description.

M.M.C., ALIAS CAPPER

Height 6ft. 1in. Weight 13st. 5lbs. Fair complexion. Wears his hair parted in the middle, but generally slightly ruffled in appearance. His criminal record, as yet incomplete and his activities are very varied. He has bigamist tendencies as a result of a sympathetic nature. Was pulled up in 1931 along with GGT for publishing seditious literature. Is definitely known to obtain considerable income from hampering the College football team. Fond of music and plays a Jew's harp passably well. Has a peculiar influence over pigs and women. Both are his playthings. It is said he is on familiar terms with every barman in the town and every rooster down the avenue. Was last seen pinching a special chocolate fish, the property of Dr Yates, who offers a reward of 2*s*. 4*d*. for information leading to his apprehension. Harmless in his lucid moments, but subject to violent outbursts of passion – all kinds.

Mac Cooper graduated from Massey in the academic year ending in 1933, which in itself was the realisation of his schoolboy ambition and a great source of pride to his family back home in Havelock. However a much greater honour was in the process of being put in train. His academic record and practical approach was well recognised, as was his ability in the sporting arena both at College level and in his representing New Zealand Universities at rugby. He also took a leading role in student affairs through his Presidency of the Students' Association and editor of *Bleat*. In other words he had been an outstanding student throughout his time at Massey. As a consequence Mac Cooper's name was put forward by the College authorities for consideration for a Rhodes Scholarship to Oxford University.

The Rhodes Trust came about as a result of the bequests of Cecil John Rhodes who made a fortune out of diamond mining in South Africa during the latter part of the nineteenth century. Better known for the colony which bore his name prior to the creation of the independent state of Zimbabwe, Rhodes, in his will, left the majority of his considerable fortune to the foundation of a unique scholarship scheme. This was to focus on the quality of the candidate's character as well as his intellectual skills. These were imperial days and the aim

Graduation at Massey
Mac standing on the right.

was to provide future leaders of the English speaking world with a broadened education and a wider horizon.

Rhodes chose Oxford for the endowment, partly no doubt because he had been there as an undergraduate, but also because he felt that the residential College system was likely to be the best environment within which the scholars might develop their abilities. Rhodes was specific in describing the qualities he required in the scholars:

> My desire being that the students who shall be elected to the scholarships shall not merely be bookworms.
> I direct that in the election of a student to a Scholarship regard shall be had to:
> - His literary and scholastic attainments
> - His fondness of and success in manly outdoor pursuits such as cricket, football and the like.
> - His qualities of manhood, truth, courage and devotion to duty; sympathy for the protection of the weak; kindliness, unselfishness and fellowship.
> - His exhibition during school days of moral force of character and of instincts to lead and to take an interest in his schoolmates for those latter attributes will be likely in afterlife to guide him to esteem the performance of public duties as his highest aim.

The original will provided funds for 52 scholarships each year from countries both within and outside the British Empire, including one for New Zealand. More recently the number of scholarships has been increased to 87, with New Zealand taking three. The administration of the scheme was placed in the hands of the Rhodes Trustees, who included, at its inception, leading luminaries and aristocrats of the day. In 1907 the estate was valued at about £3.3 million. Through skilful financial management and investment this had risen to £146 million by 1996. Although the various National Committees use broadly the same criteria for the selection of candidates as they did originally, some of the emphasis has changed in line with modern day culture. For example, sporting prowess is no longer an essential component provided individuals are able to demonstrate other physical activities which enable them to make an effective contribution to society. Women are now considered equally within the list of applicants and the appropriate amendments made to the wording in the original will.

This was the background to the famous Rhodes Scholarship to which Mac Cooper's sponsors referred. In their view, Cooper met the requirements laid down and of course were he to be successful, his selection would herald a major breakthrough for Massey College itself. It would at a stroke deliver some much needed credibility to their cause and publicise to one and all that Massey really was able to produce men of calibre equal to the best from any university in New Zealand.

It seems, in the event, that fortune smiled somewhat on Mac Cooper at the time of his application. Some may even say that he was fortunate in being selected, although the evidence to hand is purely anecdotal. The author James McNeish, who is an expert on the subject of New Zealand Rhodes Scholars, maintains that in 1932 John Mulgan, who had just graduated from Auckland University, was the obvious choice for the Scholarship. However Lord Bledisloe, Governor General of New Zealand and Chairman of the Selection Committee, felt that at nineteen, Mulgan was too young to get the best out of the opportunity and that he should return the following year.

Mulgan, however, then became involved in some political criticism of the university authorities and took the side of the students in various publications, thus alienating himself from people who might have been expected to support his second application for a scholarship. In fact his name was not put forward by the University the following year and, without their support, there was no way that the Selection Committee could reconsider his application. Bledisloe, who had engineered the delay in the first place, was furious and refused to put forward a candidate at all in 1933. The University did not relent in 1934 either so that Mulgan was forced to borrow some money from his father and went on to Oxford independently.

To make up for lost ground therefore the Selection Committee were able to put forward three candidates in 1934 instead of the usual one. According to McNeish, the first two were Norman Davies of Dunedin, who eventually became a professor at Oxford, and Ian Milner, who went on to occupy a chair at the University of Prague. The third was Mac Cooper from Massey, their first Rhodes Scholar. McNeish maintains that had not the Mulgan affair occurred Cooper would not have been awarded a Scholarship.

The other piece of anecdotal evidence and entirely in character with the applicant concerns the principal subject discussed at Cooper's interview with the Selection Committee. It is reputed that Mac Cooper, being well aware of Lord Bledisloe's interest in agriculture and his support of Massey, swatted up his special interests in farming. Apparently pigs featured prominently in this research and as this was also a particular interest of Cooper's, he determined to steer the discussion in that direction. The story goes that virtually the whole interview was taken up with a detailed debate on the various nutritional requirements of the said animal, much to the surprise and bemusement of the rest of the Committee. Whether true or not, the outcome of the interview saw Mac Cooper as one of the successful applicants for the highly prized Rhodes Scholarship in 1934.

Needless to say there was much celebration both within the College and back in his hometown. He became a celebrity overnight and his reputation approached that of the 'Boy's Own All-Rounder' to whom all prospective mothers in law aspired for their daughters. What exactly he was planning to do at Oxford had really not been considered. Agriculture was not an obvious subject in the 'City of dreaming spires'. Cambridge would have been infinitely preferable, especially for an animal production man like Cooper. The world famous animal scientist John Hammond was working there and to have been under his wing would have been a tremendous opportunity. But the immediate priority lay in earning some money to help fund his 12,000-mile voyage to the other side of the world. The scholarship paid for the fees and went towards living expenses, but it did not in those days cover the fare to Oxford.

Massey was happy to employ him as a part time lecturer after he graduated and he also undertook some warden duties within one of the residential blocks at the College. According to Ian Campbell who was a student at Massey at the time and later returned after completing his postgraduate work in America to the Chair of Animal Husbandry, Mac was in charge of the students on the farms and the driving force behind the hay and silage making operations. He also continued to play rugby and cricket for the College and was instrumental in mobilising student labour to build a swimming pool in the grounds. A local newspaper article with the headline 'Enterprise of Students' relates the commencement of the work. It reads: 'Not only the students, but the paid staff and the Principal of the College (Professor G.S. Peren) are participating in the heavy manual labour associated with their task. They are very capably demonstrating what may be accomplished by initiative and self-reliance at a time when the straightened state of finances precludes State assistance being given.' The swimming pool was eventually opened in March 1935 and served the community extremely well until the base was cracked as a result of an earthquake and the site filled in for a car park in later years.

Following the award of the Rhodes Scholarship there followed a long series of correspondence with the authorities as to the appropriate course of study that Mac Cooper should undertake when he arrived in Oxford at the beginning of the Michaelmas term in October 1934. It seems that the residing Dons were unimpressed with the news that one of the forthcoming Scholars was to be an agriculturalist from New Zealand. A letter from the Warden of Rhodes House to Lord Lothian who was the Secretary of the Rhodes Trust, dated 31 January 1934, illustrates the point:

My dear Lothian
    New Z. becomes curiouser and curiouser. As one of their three R. scholars for next year, they have selected a man who has taken a degree in Agricultural Science, and is coming to Oxford to take the Diploma in Rural Economy.
    His educational record shows hardly any 'liberal' subject at all. His speciality is pig feeding. He seems a good and intelligent fellow, but, if

unselected candidates like Strang and Mulgan are as good as they are said to be, it seems odd to send a man all the way from New Zealand to study purely utilitarian subjects. I don't know what College will want a Diploma candidate.

It may be coincidence, and it may be gossip, but I am told that Lord Bledisloe is deeply interested in pigs and their diet.

Yrs sincerely

C.K. Allen

Mac Cooper did, however, manage to persuade the powers that be that his choice of University College was in their best interest. Perhaps Lord Bledisloe had been used as a reference as he himself had been a Univ. man. There followed throughout the next six months a prolonged correspondence about whether he should read for the Diploma in Rural Economy or for a degree in Agriculture. Neither his College nor the Rhodes Trust liked the idea of a Diploma, as they were not convinced that its content would be sufficiently challenging for able students.

Mac Cooper aged 23 – Rhodes Scholar.

In reply to a question from the Warden, C.S. Orwin, Director of the Agricultural Economics Research Unit, wrote that he agreed about the unsuitability of the Diploma, but also considered that the degree course would simply cover the same ground as that already experienced at Massey. He ended somewhat in desperation: 'Alternatively, do you think the Honours School of Geography would be of any use to him?'

These concerns were passed on to Mac in April and he was asked to discuss the matter with his professorial colleagues at Massey. Meanwhile a further exchange took place between the Warden and the Dean of University College, with whom he was on Christian name terms. The Warden points out that discussions with Orwin make clear that the Diploma course would be unsuitable as well as insufficient and he also doubts whether the School of Agriculture would be any better, as he would not get much more out of the course than at Massey.

He goes on to say:

In short there is very little we can teach him about pig feeding, bran mashing or mangold wurzling. It looks as if he ought to find some subject for an original contribution to knowledge leading to a B.Sc. or even D.Phil. e.g. Moral and Sociological Aspects of Bacon Rind? I had better retract my letter don't you think exhorting him to the School of Agriculture?

The Dean's hand-written reply in a similar vein was as follows:

I agree. Let us urge the good Cooper to plunge into potato smut or leaf mould in wheat with special reference of course, to the League of Nations and Professor Zimmerman. That will be much the best. I will advise the Master accordingly.

Whilst on the surface this exchange and others between University colleagues may seem rather childish and indicative of intellectual snobbery, they show a concern to get it right as well as illustrating their amused lack of understanding about matters agricultural. Eventually, after the involvement of Professors Peren and Riddet, Cooper replied to the Warden that he wished to proceed with the course in Rural Economy and then proceed to a higher degree or research.

The type of thesis I have in mind involves a detailed knowledge of both British and New Zealand agricultural organisation and I feel the Diploma course would give very good foundation knowledge of British agriculture.

After discussion with various people in New Zealand, a subject for a D.Phil. research project was agreed with the title: 'Nature of Demand for Dairy Produce in the UK with special reference to NZ supplies.' Perhaps in view of the nature of the chosen subject, there was an influence emanating from the Department of Agriculture in New Zealand. Such information would be of value in assessing future butter and cheese exports to Britain. In fact, it became clear as time progressed that Mac Cooper had been given the promise of a job back in New Zealand after he finished at Oxford with the newly established Department of Scientific and Industrial Research based in Wellington. So it is hardly surprising that the subject chosen for his research was influenced in some measure by the Ministry boffins back home. Whether there was any financial incentive to this arrangement during his time as a student is not known, but it is clear that he was committed to return to New Zealand after he had completed his studies.

In any event the proposal to follow the Diploma with a higher degree seemed to satisfy the sensitivities at both Rhodes House and University College so the scene was now set for Mac Cooper's enrolment at Oxford in the October of 1934.

4

OXFORD

In the July 1935 edition of *The Bleat*, an article was published entitled 'Pot Pourri'. It was contributed by Mac Cooper and describes in some detail his first six months as a student in Oxford.

There is no need yet for me to introduce myself, I still feel so much one of you. Cheers, my old friends; I am going to tell you a few of my experiences and to give you snatches from my unwritten diaries.

THE VOYAGE

I wasn't seasick. The old Viking blood of the Coopers rose to take charge of my stomach. It was a voyage with little incident apart from a night ashore at Panama and the traversing of the magnificent canal. We arrived at Panama in the late afternoon after sighting land, the first land for three long weeks, early that morning. The excitement on board was intense. Even those elderly maiden ladies who composed such a large proportion of our company showed unmistakable signs of joy. It was a luxury to walk on land, to drive in an open car with an attentive black chauffeur, feeling like regular tourists. We saw some good cabaret shows that night and when I say shows I am not fooling. Of course Panama was very warm so we cannot blame the young ladies for wearing so little. The canal itself is wonderful, a monument to the advances of engineering, worked with a quiet efficiency that makes one suspect there is a Scotsman in charge.

Thirty-three days from Auckland we saw our first sight of England, the Scilly Isles; by nightfall we could see Michael's Mount. It was the England that I had so long dreamed about. Some day before I return to dear old NZ, I am going to climb Michael's Mount. By breakfast time we were abreast of Eastbourne and Beachy Head. That was my first sample of the white chalk cliffs of Southern England. I cannot describe the excitement of that day, the fresh points of interest cropping up all the time, the exclamations of joy from returning travellers as they recognised familiar landmarks and the novelties we new arrivals were experiencing. During the morning we picked up the Channel pilot who took us past the White Cliffs of Dover to the Thames Estuary. By six o'clock we were abreast of Tilbury, but it was not till midnight that we reached our safe anchorage in the heart of Dockland. Next morning we made our farewells to our shipboard friends and entered mighty London, the city of extremes, the city of adventures, of hopes and of miseries. I had friends to meet me and I think it was just as well because I was a very bewildered colonial, far from his natural field of turnips.

OXFORD
I arrived at Oxford just as darkness had fallen. I found a taxi and very proudly said, 'University College.' I felt that I could not then be so familiar as to call it 'Univ'. I recognised the College a hundred yards away from it, thanks to Dr Watt's faithful description. I entered the grey walls across the worn steps with a thrill of pride. The Porter took me to my rooms that did my heart good – a large carpeted sitting-room furnished with three easy chairs, a desk and desk chair, two tables, a sideboard and a bookcase. My bedroom did not disappoint me either. I made the acquaintance of my staircase scout who made me quite at home. These College scouts are a great class of men and as loyal to their College as they can be.

That night I had my first dinner in Hall. I sat at a long table amid a noisy crowd of reserved Englishmen who said not a word to me. I went back to my rooms a very lonely man. My letter writing to the accompaniment of the interminable ringing of bells, so characteristic of Oxford, was interrupted at ten o'clock by the arrival of a visitor, a Canadian who had rooms next to me.

What a joy it was to experience colonial friendliness again! The next day I mustered my declining knowledge to enrol in the traditional manner, collected my first Rhodes instalment, ninety real English pounds and opened a bank account.

Another interesting formality that day was the Matriculation ceremony. The eighty Univ. freshmen, all in academic dress, moved in disorderly order to the Clarendon buildings where the Bedel presented us with a copy of the Statutes. We signed our names and the Chancellor muttered a Latin incantation over our heads, bowed, raised his cap, bowed again to place us *in statu pupillaris*. Next week rugby commenced; in the earnest struggle for a Blue, I forgot my loneliness.

'VARSITY

There are twenty-three men's and four women's colleges in the University with approximately four thousand students either in residence or in lodgings. Oxford is heterogeneous. The students' consuming interests range from classics to hunting. They are drawn from every part of the world and from nearly every walk of life; men with black curly hair and with faces to match; Asiatics and representatives from nearly every European country. That is perhaps why Oxford offers so much to her students. It has an infinite variety in an Old World setting of buildings and tradition. The fine old buildings cannot but appeal to the colonial. I do not tire of my Sunday afternoon jaunts of exploration in company with my two refreshing Canadian friends who insist on posing as American tourists and making offensive remarks about the college they are inspecting for the benefit of any undergrads. who may be within hearing. They look tough men so they have no fears.

An undergrad. is not allowed to enter a drinking house except in pursuit of food and only then if the Proctors license it. But this does not mean that the undergrad. is temperate. Ale and wines are available in large quantities at the College stores while most men keep a bottle or two in their rooms, just in case there is a drought. But drinking is much more of a regular feature than we are accustomed to in New Zealand. A glass of ale, bread and cheese is the traditional Oxford lunch. There are excesses though, particularly during periods of great rejoicing as are occasioned by such events as winning the Inter-College Rugby Cup or being the Head of the River.

Univ. won the Cup this season. The competition is run on knockout lines and played in a spirit comparable with that displayed in the Inter-Faculty Competition back in New Zealand. The night of the final we had a 'Bump supper'. Champagne and assorted drinks flowed merrily, the ramparts of the College were adorned with articles of bedroom pottery and foolish virgins who had imbibed well but not too wisely were led to their beds. Certain of the stronger vessels lasted the night through and attended roll call in the morning still in dinner jackets.

There is a quaint custom in Hall of sending along a pint of ale to a friend per medium of the beer scout. One night after a 'Varsity match, another Univ. member of the team and I brought a couple of team-mates into Hall as guests for dinner. Between us we had forty-one pints of beer, mostly in silver tankards, some of them over a hundred years old, sent along with compliments of admirers. They made such a noble gleaming array about us that I almost regretted my teetotalism. It's a great life if you don't weaken.

There can be little doubt on reading this that Mac Cooper found England

and particularly Oxford, its traditions and its vibrancy extremely invigorating. At first he seemed rather in awe of his fellow students with their rather snobbish approach to someone whom they categorised as a Colonial. Despite his track record he remained quite shy by nature and unwilling to push himself forward until he found his feet and self-confidence. It didn't take long. He was selected to play in the 'Varsity match as a freshman, an honour which at the time, he would have undoubtedly rated as the highlight of his career so far. He recalls the experience in the 'Pot Pourri' article:

> I will never forget the day I was asked to play for 'Varsity against the Tabs (the nick-name given to the Cambridge or Cantab. team). It was a Sunday and I had made the day sublime by going to church in the morning. I had eaten a sparing lunch and was taking a walk along the High by myself, when a car drew up by the kerb. The Captain of the OURFC, Derrick Lorraine, hailed me and gave me a letter. At the same time he extended a hand of congratulation and told me the letter was an invitation to represent the 'Varsity on December 11th. An invitation!
>
> I rushed back to Univ., my feet scarcely touching the pavement, seeking my friends to tell them the good news. I couldn't find any of them, so all I could do was wander around aimlessly, looking very pleased with myself. I wasn't teetotal that night. I was the Captain's guest at his College and drank from three-handled porcelain beer pots that held a quart. During the evening I entertained with a vivid rendering of the Massey haka. From then on till the match the tension gradually increased. We went down two days early and spent five days at Eastbourne taking the sea air and getting more and more nervous. The day before the match we went to Oaklands Park where we spent a sleepless night in the luxury hotel that none of us were in a fit state to appreciate. A fleet of Rolls Royces bedecked with navy blue carried us to Twickenham through the tremendous match traffic. It was a proud moment when I passed under the gaze of thousands from the car to the changing rooms.
>
> I was glad to get on to the field. As soon as I touched the ball and as soon as I had downed my first Tab. I forgot the fifty thousand spectators. I won't say anything about the match except, how I did long for Bill Pease, Jack Waters and Dan Pierce to be beside me.
>
> Next year it will be different. Already the prospective pack is responding to New Zealand tactics and we are playing 3-4-1 and adopting a few of those vices that are so essential.

Mac Cooper took rugby football extremely seriously. In those days to get a 'Blue' at either rowing or rugby was probably more important than getting a first class degree. Indeed many of the Colleges, in their student selection, often overlooked academic shortcomings and enrolled men who had a sporting pedigree from school or from an overseas institution. With the sporting criteria laid down for Rhodes Scholars, it is hardly surprising that a high proportion of them achieved distinction in this respect. This could well have been a factor in Mac's successful application to University College, overriding the debate about the course of study he was to undertake.

The 'Varsity match in 1934 was the first of three successive occasions he played for Oxford against Cambridge. In his article for *Bleat*, Mac did not elaborate on the match or the result. In fact it was little short of a rout, with Cambridge winning by 28 points to 3 and that was when tries were only worth 3 points.

Howard Marshall in his book *Oxford v. Cambridge The Story of the University Rugby Match*, describes the carnage. 'Cambridge cut loose at Twickenham and picked Oxford up and threw them down and tore them to pieces and beat them by two goals, a penalty goal and four tries to a drop goal. A notable victory by a great team. The swing of the pendulum, also, for this was the first Cambridge win since 1928. The post-mortem was complete. Lobbed passes from the scrummage, weak tackling on the wings and the corpse of Oxford rugby was plain for all to see.' No wonder Mac didn't elaborate!

By the following year Cooper was Secretary of the University Rugby Club, although he actually captained the side in the 'Varsity match, as the appointed captain, Ken Jackson, was injured. The match ended in a pointless draw for the first time since 1892. Marshall's account of the match (he was an Oxford man) described it as an equitable result although he found it 'not easy to write with temperate judgement of such a grand, tear-away game'. He goes on:

> Cambridge were not quite as strong behind the scrummage as they had been in the previous year. They still had Wooler, Fyfe and Cliff Jones, a dangerous trio in all conscience, but Candler and Johnston had gone down. Oxford had one recruit, Obolensky the Russian Prince, of outstanding quality, while in the centre that most capable footballer and all-rounder M.M. Walford had joined the side. Obolensky on the wing was an exciting player, as those who saw his remarkable try against the All Blacks for England in 1936 will remember.
>
> He was exceptionally fast, but his great asset was his deceptive change of speed in full flight and his ability to accelerate past an opponent as if he had slipped into top gear. With head flung back and long fair hair streaming in the wind, he was one of those rare players who impress themselves on the memory, and his death in the 1939-45 war saddened the Rugby world.

Obolensky indeed became a legend and this try to which Marshall refers, remains part of the folklore of English rugby even to this day. At the time he was an adopted Oxford hero with his every movement recorded. This is well illustrated by a local Oxford newspaper cutting of the time with the headline:

REGRETTABLE CRACK
Ice is no respecter of persons or records. On two occasions last week the ice refused to support Prince Obolensky in the state to which he has become accustomed, and he made the acquaintance of what lay below the surface.

Cooper was duly elected captain of the University Rugby Club in his own right the following year. He made some controversial changes. In previous years, 'Old Blues' were exempt from the final trials to get into the first XV. In typical style he swept this privilege away so that they had to fight for their places with the up-and-coming youngsters. He also designed his own personal invitation card to those who were finally selected to play in the 'Varsity match.

> *O.U.R.F.C.*
> *Dear.............*
> *You are wanted by your Captain to assist in*
> *TAB SLOGGING on Tuesday, December 8th, 1936.*
> *I trust that this invitation will not interfere with your*
> *social arrangements.   Your number is.........*
> *   Congratulations!*
> *   Mac.*

Despite all efforts however, Oxford were to lose the 1936 match by the agonising margin of just one point. Marshall, waxes eloquent about the game and before it started was confident that this year would see a return to winning ways.

The weather was wet and the greasy ball favoured the more powerful Oxford pack. Surely, we thought, they would control the game and fast following and Grieve's kicking would do the rest. In such conditions the Cambridge midfield attack, justly praised and greatly feared by Oxford pessimists, would be powerless. The gods were with Oxford. And gentlemen in Dark Blue ties took luncheon in good heart. They would have choked over the cheese had they known what was to follow.

The rain dripped dismally from heavy skies, but our anticipation was weatherproof, and when Cooper led the Oxford team out of the dark tunnel the continuous baying of the crowd began. Rain, remember, does not spoil the game, but it alters it. And the better side is the one which adapts itself to the circumstances most successfully. They scored two tries to one; there can be no doubt about that. A draw would have been sentimentally satisfying, perhaps, but the purely moral story, in which justice triumphs and virtue is rewarded, demands a victory for Cambridge.

Mac Cooper's rugby career whilst in England at this time extended well beyond the University matches. In 1936 he was selected to play for Scotland against Wales at Murrayfield on 1 February. Clearly the process of identifying ethnic ancestors in order to qualify to play for the home nations is no recent phenomenon of the professional game and Mac may well have been one of the first New Zealanders to do so. There is certainly no doubt about the origins of his maternal grandfather, Malcolm McGregor, although in those days, it is unlikely that there were investigative journalists checking on the relevant birth certificates.

He again wrote of his impressions of this personal milestone for the benefit of the readers of *Bleat* back home:

I will never forget my first international because it was such a tremendous occasion. Three years ago I would have laughed if anyone suggested that I should be playing for Scotland, but the improbable came true. I left Oxford on the Friday to join the Flying Scotsman at King's Cross. At Grantham, Cliff Jones and Wilfrid Wooler joined the train, so I had illustrious company for the rest of the journey.

At the hotel I met a number of the Scottish team who had arrived from a distance. We went to a picture and then after a glass of hot Horlicks (to guard against night starvation) retired to bed at a reasonably early hour. When we came down to breakfast the lounges were crowded mostly by Welshmen, and all of them were talking about the game to follow. I was starting to get excited. After breakfast I went for a stroll along Princes Street, which presented a marvellous sight. It was absolutely crowded with a slow moving cheerful mob that was killing time before going along to Murrayfield. Over fifteen thousand Welshmen were in Edinburgh that day. They came up in forty trains during the night. Practically all of them wore scarlet berets and some of them even went to the extent of wearing red trousers and coats. Leeks were very much in evidence, while a number carried pans to do service as kettledrums. There is something peculiarly Welsh about a saucepan. Llanelli (pronounced Clanethly) is a centre for their manufacture. The Rugby Club has adopted the saucepan as their symbol. When the Varsity played Llanelli last Easter I was presented with a miniature saucepan and the All Blacks received a similar presentation. Llanelli have a rousing club song, which is called '*Sospan Faph*' or little saucepan. The theme of the song is that the saucepan should be kept entire – a loss means that there has been a hole put in the saucepan. Mine had no flaw whatsoever, but those the All Blacks received figuratively had the bottoms knocked out.

There is something rather splendid about the way these Welshmen follow their team. Most of them are miners and factory workers, living on what is practically a starvation wage. They form supporters clubs and contribute 6*d.* per week for two years and this gives them enough to make the Edinburgh trip. For many, these trips are the only ones they have away from their native district with its succession of slag heaps and general air of depression. They make the most of their excursion, and are generally up to all sorts of tricks. Two years ago a party of them entered Murrayfield in the wee small hours and painted the goal posts red. They tried to do it this year, but the canny S.R.U. had watchmen present. Another favourite trick is to come out onto the field just before the game to try and hang a leek over the crossbar. This gives the policemen something to do and gives the waiting crowd something to laugh at.

There were dozens of men in the street selling badges and colours. I managed to resist a number of them, but finally one persuasive blighter bullied me into buying a thistle that wasn't tied with McGregor tartan. I wore it back to the hotel.

After lunch we had our team talk and then went by bus through the crowded streets to Murrayfield. I forgot one important episode. During lunch I was handed my Scottish cap, inside up in a large brown cardboard box. At the earliest opportunity I furtively escaped to my room to examine my prize. I couldn't resist trying it on. I have almost come to the conclusion that I have a feminine streak, but I did not practise my vanity long. On the way to the ground we sang incredibly vulgar songs. The broad Scots accent seems to take the edge off the vice in some indefinable way. For just such occasions I have an accent too, but people tell me that it is a New Zealand one.

Once in the dressing room my nervousness left me. I think that was partly on account of my pansy white trousers which are a real thrill, They have such a wonderful billowing sweep and a very nice pair of pockets, where you can put your hands after finishing your orange at half-time. I was very pleased too, to receive number thirteen jersey. I have worn the devil's number continuously for the Varsity and for my College since I have been over.

After the photographs we trotted out onto the field and the band struck up 'God save the King'. It was just after King George's death and we were wearing black armlets. The mighty crowd of seventy thousand rose as a body and sang the Anthem, as I have never heard it sung before. I felt very small but very proud standing out on that pitch with my fellow players. It made me think a lot.

The game started immediately. It was hard and fast from the very beginning. In the first scrum I had a fist pushed into my face, but I got a good one back so he didn't hurt me. The Welshmen are really tough to play against. They fight and kick and obstruct for all they are worth, but for all that they are very good chaps – after the game. It was rather strange for me in that game. On my own side they were mostly all broad Scots from the Border, and they talked such an accent that they don't even understand one another. On the other side you had the Welsh language which is a trifle more difficult to comprehend. I dared not say anything in case people suspected that I was an outsider.

We held them in front, but Wooler and Cliff Jones were too much for us and they scored the deciding points after a magnificent movement. The Welsh spectators sang 'Land of my Fathers' and my word can they sing. Our own supporters were too self-conscious to respond on our behalf. When 'no-side' blew, the crowd swarmed onto the ground. I got caught by a mob of small boys seeking autographs and was rescued only in the nick of time by a burly policeman.

A good hot bath and a couple of shandies restored some of my pristine vigour. We went back to the hotel and revived the tissues a bit more before getting ready for dinner. There is nothing like a good cup of tea. Dinner was a very quiet function on account of the royal bereavement, but no one appeared to be too sober. I ate my haggis without turning a hair. I reckon, despite what some of the newspapers over here say, that I really have some Scottish qualifications when I can do that sort of thing.

Although Mac omits to include the score in this article, it was in fact a victory for Wales at 13 points to 3. Not many of Mac's team mates register in the memory now, but the opposition, apart from those mentioned above, included the legendary Vivian Jenkins at full-back and Hayden Tanner, the Swansea scrum-half.

Three weeks later Mac was in the Scottish team to face Ireland, again at Murrayfield. The result was somewhat similar ending in the score line of 4 points to 10 in Ireland's favour. He was also selected to play in the Calcutta Cup match against England at Twickenham, a venue every rugby player aspires to. But it was not to be, as he had to stand down because of injury. Scotland narrowly lost that match as well by 9 points to 8 and took the Wooden Spoon for 1936 – hardly a vintage year.

Another highlight of Mac's rugby career was the Oxford University match against the touring All Blacks of 1936. For obvious reasons this was a special occasion for him and the students came within an ace of a remarkable victory. With three minutes to go to full time, in what had been an excellent contest in poor weather conditions, Gilbert, the New Zealand full-back, kicked a conversion from the touchline to clinch the match by a single point at 10-9.

There were of course numerous other rugby encounters throughout his three years at Oxford. These included University tours to Ireland and Scotland and to California, where the team was based in Hollywood, a somewhat surprising venue in those days. In fact, in March 1935 Mac had occasion to

Scotland versus Ireland at Murrayfield.
22 February 1936. Mac standing third from left. (Scottish RFU)

Oxford University versus the All Blacks. 1936. Mac Cooper, the Oxford captain seated fifth from left.
G. Gilbert who kicked the winning conversion is the player in the middle row, seven from the left.
On his left is Prince Obolenski. (G. Gilbert)

write to the Warden of Rhodes House requesting an advance of £25 from his next term's allowance. This was required to meet 'the expenses entailed through the American rugger tour which have been and will be fairly substantial – otherwise I would not make this request.'

He was also invited to play for the Barbarians, on their Easter tour of South Wales in 1937: an honour highly regarded by rugby players the world over. He played on the winning side in three of the games, scoring in one of them. In the final game against Cardiff, the Barbarians lost 3 points to 7, ending a series of 22 consecutive victories and recording their first defeat in South Wales since 1930.

In February 1936, the University student's magazine published an article on Mac Cooper under the title 'Isis Idol'. To be the subject of such an accolade points to his 'super star' image amongst his peers. The article states that Mac 'has proved himself an ideal rugby skipper on and off the field.' It goes on to describe his background in New Zealand and his track record at Massey, culminating in his Rhodes Scholarship. Much of the content is taken up with his rugby prowess, but one interesting paragraph reveals his obsession about the incorrect spelling of his initials, which drove him into frenzy. 'Newspapers have indefatigably referred to him as McGraw Copper, McGowan Cooper and Mc-Cooper and letters have been known to have even further variations. But McG represents yet another famous Scots clan, as he tells us in his quaint old English accent.' The article ends: 'Mac is destined for a post in the New Zealand Agricultural Department, and with the enthusiasm for his subject and his charm of manner, there is no doubt that Mac will rise to the same heights in his profession as those to which he has risen in Oxford Rugby.'

There is limited information as to any detail about Mac Cooper's progress once his academic programme had been agreed. He joined the School of Rural Economy where he came under the tutelage of Professor Scott-Watson and Dr Keith Murray. He found the work rather theoretical and uninspiring, as he had previously been closely involved with animal husbandry work at Massey, both in the lecture room and on the farms. He would certainly have preferred cows and pigs rather than have to grapple with the law of diminishing marginal returns. Some years later in a CV which he wrote when applying for the job at Newcastle, he stated that he was disappointed to find that there was no opportunity to study animal production at Oxford. Given the choice he would have much preferred to work at Cambridge under John Hammond at the Animal Research Institute. Hammond was a world authority on the physiology of reproduction, growth and meat quality of farm animals and ambitions young researchers from far and wide were anxious to come and work under his guidance. Indeed one of these was to be McMeekan, who had recently been given leave of absence from Massey to come to Britain to do just that.

Cooper had good reason therefore to visit his friend when he arrived in

The 'Sporting Idol' at Oxford

Cambridge and wrote in a private letter to Gordon McLaughlan:

> Though I was at Oxford I saw a lot of Mac at Cambridge because I used to come over and help him with his dissection in his classic plane of nutrition study on growth in pigs, which instantly gave him an international reputation. Apart from the pleasure of helping, it gave me a chance of sitting alongside John Hammond who took his full share in the exacting task of dividing 145 lb. carcasses into bone, fat, and muscle, with the provision that bones had to be so clean that they would not even attract a blowfly.
>
> A typical day would start at 6 am with the slaughter of two pigs with everything measured down to the last drop of blood. By 9 am they would be in the laboratory ready for dissecting by a team which always included Hammond and his technician Pike and several others who had been press-ganged into action, including Pallson from Iceland, Verges from Argentina and Messerschmidt from Germany, all fellow students under Hammond. It was a League of Nations group and the general discussion reflected the breadth of their backgrounds.
>
> We would work on until 9.30 pm in time for a couple of pints before closing time with only time for two short breaks for something to eat. I remember protesting vigorously when pork pies were produced after fifteen hours of cutting up pigs.

Back at Oxford the Rural Economy course ground on to the satisfaction of his tutors, although it seems that Cooper's priorities were elsewhere, primarily on the rugby field. His first term report from the Master of University College to the Warden of Rhodes House states that 'not only has he got a Blue but reports from Watson and Murray are outstandingly good.' The Master adds somewhat patronisingly, 'He has settled down surprisingly quickly and we all like his horse-like charm.'

However communications between the academics and College administrators seemed to deteriorate thereafter as in the following year's report to the Warden, the Master says: 'We have every reason to be satisfied with the conduct and industry of Mr. Cooper. This opinion is formed, not from information about his actual work, on which nobody seems able furnish us with a definite report, but from our own observation of the man, who seems a very sound, level-headed and a serious person.' In any event Mac was successful in applying for a third year extension to his scholarship in order to pursue the research project agreed at the start of his time at Oxford. This was in pursuit of a D.Phil., which would have been a significant achievement and added considerable weight to his academic track record. The title of his thesis was 'The Nature of Demand for Dairy Produce in the United Kingdom with Special Reference to New Zealand Supplies.' No record exists of this dissertation, indeed if it ever materialised. A paper did however appear in the *New Zealand Journal of Science and Technology* in March 1937 under the title 'The Demand for Cheese in the City of Oxford', the author being M.M. Cooper. This paper could not be described as being of classical importance and it is difficult to see any of the Cooper enthusiasm emerging from the house to house survey carried out by the Agricultural Economics Research Institute.

The summary of the paper tells all:

> This article presents an analysis of answers given to a series of questions relating to cheese consumption in the city of Oxford. The survey, which was conducted in May and June 1936, covered 587 households, of which 459 regularly purchased cheese and 87 purchased it occasionally. The type of cheese in widest demand was Cheddar, but processed and fancy cheeses occupied relatively only a minor place in consumer demand. Processed cheese had an important function as an occasional purchase either by households not buying cheese regularly or households that bought as a change from standard bulk types.

Back at the ranch, so to speak, his termly reports continued in repetitive vein. For example, at the end of the Trinity term in 1936, the Master writes 'Mr. Cooper has been elected President of the College and I hope that this, taken in conjunction with his other activities, will not prove to be an impediment to his getting his Doctorate in Philosophy. We find him very straightforward, but about his actual work we can give you no detailed information. He is one of the pillars of this house, and we are very glad that the Rhodes Trustees have been able to continue his scholarship for a third year.'

In this final year Mac Cooper applied his energies to a wide range of extra-mural activities. It is probable that he foresaw the impending problems with the sheer size of his original research project and decided that at least he would put his heart into things which both interested him and from which he could gain experience about people and life. As well as College President and his rugby commitments, he became President of the Raleigh Club (concerned with the welfare of overseas students) and Vincents (an invited club for all Oxford 'Blues'). He was involved with the setting up and organisation of unemployment camps in the locality as the recession was still in full swing with poverty and unemployment remaining at very high levels. He was also a member of a committee and joint initiator of a University Malnutrition Scheme, perhaps directly linked to the same problem. Finally, and perhaps an unusual interest at that time, was his membership of the College Shakespeare Club, a dining and speaker Society. Thereafter his love of Shakespeare persisted throughout his life.

Social life at Oxford must also have been a constant round of parties and events and with Mac Cooper's sporting reputation, he was bound to have been

Social life at Oxford was an enjoyable experience.

a popular and regularly invited guest. It was in this connection that he met Hilary Mathews, whose elder sister Barbara later married Derrick Lorraine, the Oxford University rugby captain in Mac's first year. Hilary's father Wilfrid came from a well-known Oxford family and had had the good fortune to be friendly with Alfred Morris, the inventor of the famous car which bore his name. Wilfrid was in the family insurance business and quickly spotted the possibility of diversifying into motor insurance. This had its ups and downs but became very profitable after the war, when most of the cars were exported abroad. Wilfrid's young wife Florence was something of a socialite with artistic talent, coupled with an inclination to entertain lavishly. As the recession deepened they had to move from fashionable Boars Hill, on the outskirts of Oxford, back into the city and virtually live above the office until their fortunes improved in line with the insurance business.

Mac's friendship with Hilary blossomed into romance and in July 1937 he wrote to Dr Allen, at Rhodes House, informing him of his forthcoming marriage and explained that there was a problem in finding a suitable venue for the reception. He wondered whether there would be any possibility of holding the celebration in one of the rooms at Rhodes House. He adds, 'I hope you do not feel I am being presumptuous in making this enquiry for such a personal favour. You will understand my feelings – Rhodes House has meant so very much in

Three consecutive Oxford University Rugby captains.
Mac 1936, Derrick Lorraine 1934 and Ken Jackson 1935.

my Oxford career that I felt any possibility would justify my asking you.'
Unfortunately Dr. Allen was unable to help. He replied:

> I should very much have liked to offer you hospitality here, and should
> have felt it entirely appropriate, but there are difficulties, which I don't
> quite see my way to overcome. The chief is the question of precedent.
> Every year, one or two Rhodes Scholars get married as soon as they have
> finished their course, and not infrequently the wedding takes place in or
> near Oxford. It would not be suitable in all cases to make Rhodes House
> available (in fact I have already refused on one occasion) and, on the
> other hand, it would be invidious to refuse permission in one case and to
> grant it in another. On the whole, I am afraid it is wisest to adhere to the
> rule that Rhodes House is a public building at the disposal of the
> University, and is not available for purely private functions, except such
> as we give ourselves.

Allen did in fact say that had they not being going away at the time of the
wedding he would have suggested that the reception be held in his private
house. From this it is clear that the Warden had a high opinion of Mac Cooper
by the end of his time at Oxford. Indeed his final report to the Rhodes Selection
Committee in 1937 was as follows:

M.McG. Cooper (University)
Unquestionably one of the most successful Rhodes Scholars of recent
years. Not naturally a very clever man, he has established himself in a
unique position by sheer force of character and determination. The list
of his various activities and interests in Oxford is extraordinary, and he
has made a mark on all of them. He leaves the highest opinions behind
him, and it is extremely satisfactory to feel that he is entering the service
of his own country, in which he will certainly carry out the principles and
aims of the Rhodes Scholarship. It would be impossible to find a man
with a higher sense of duty.

This is a far cry from the Warden's correspondence three years earlier.

Mac Cooper and Hilary Mathews were married on 11 September 1937 at St.
Peter's in the East Church in Oxford. For obvious reasons none of his family
or friends back in New Zealand could attend the wedding; however the occasion
was one of Oxford's marriages of the year and featured prominently in the
local press. The best man was Ken Jackson, the 1935 University rugby captain,
and the bridesmaids were Hilary's two sisters Barbara Lorraine and Sonia
Mathews. The reception was held at 10 South Parks Road and by all accounts
it was a splendid occasion.

Oxford had, therefore, lived up to all expectations with the important exception of intellectual challenge for an agriculturist interested in animal production. Mac did not submit his D.Phil. thesis and thus ended up with a Diploma in Rural Economy and a B.Litt., neither of which added significantly to his academic record. In terms of reputation within the agricultural industry, Oxford at that time was little more than a backwater. Rural Economy and Economics were either an excuse for the sons of the landed gentry to enjoy themselves as undergraduates at father's expense, or for budding civil servants destined for the Ministry of Agriculture. Had the Rhodes Scholarship been at Cambridge, there is little doubt that Cooper would have excelled under Hammond's guidance with the possible result that his career would have headed in a research direction rather than back into teaching.

But academic challenge was only part of the story. Cooper became an icon on the sports field and in many other directions he showed initiative and a willingness to help people less fortunate than himself. From a diffident start, he grew in self-confidence and leadership became second nature to him. He led from the front and by example. In other words he was the archetypal Rhodes

Mac and Hilary on their wedding day 11 September 1937. St Peters in the East, Oxford.

Scholar and true to the objectives of the scholarship he was destined to return home and serve his country in whatever way was necessary. Finally, he may have left Oxford without his doctorate, but he had acquired an infinitely more worthwhile prize – a vivacious and attractive wife – with whom he would spend his life for the next 52 years.

So after a brief tour of some of the British research stations for his new employers in New Zealand, the Department of Scientific and Industrial Research, Mac and Hilary Cooper set out on their five week sea voyage to New Zealand on the *Rangitiki*, at the end of September 1937. For Mac it was the homecoming he had been looking forward to for months but for Hilary, at the tender age of twenty, it was a daunting journey. Away from her friends and family, to a strange land down under where she knew nobody; where their night was our day; where you travelled north to warmer climes and even Christmas was in the middle of summer.

5

THE WAR YEARS

After what was really an extended honeymoon aboard ship and twelve thousand miles of ocean, the Coopers arrived in Auckland in late October 1937 to be met by most of Mac's family. They returned to Hastings to what amounted to a hero's welcome. To have been to one of the world's leading seats of learning and, more importantly in New Zealand eyes, to have captained the University rugby side as well as play for Scotland filled the newspapers. Here, in the modern idiom, was the 'local boy done good'. There was much speculation as to which rugby club he would favour with his patronage. One report, with an oblique shot at a disappointing All Black tour to the UK the previous year, quoted:

It is to be hoped that Mr Cooper passes on his knowledge of English, Welsh, Scottish and Irish scrummaging to some New Zealand Club.
'I've brought my rugger boots with me,' he confided, 'but I don't know yet whether I will play. I will be stationed in Wellington, and if I do play I'll probably play for the Wellington Club.'

But the time had come to start earning a living and within a few weeks he set up home in Wellington and each morning was at his desk in the headquarters of DSIR. In December he wrote to Dr Allen at Oxford to tell him of their safe arrival and that his new job was progressing well.

I like my work very well despite the fact that it is in an embryonic stage so far as I am concerned. It promises to develop and there are prospects that I shall be Co-ordinating Secretary of a Bureau of Animal Research which is being constituted. I do not know whether I want the position or not because acceptance of it would mean that I would not be returning to England within two years. For my wife's sake I would regret that.

This post did not materialise, or if it did, Cooper didn't get it. He was promoted within a few months to Assistant Professional Officer in succession to W.M. Hamilton, who had been transferred to the London office as Scientific Liaison Officer for DSIR. Hamilton had been a contemporary of Cooper's at Massey and later returned to take over as Secretary or Chief Executive of the Department. Early in the New Year some correspondence was exchanged between Cockayne, the Director General of the Department of Agriculture, and the Secretary of DSIR concerning McMeekan who at this stage was still working with Hammond in Cambridge. The letter concerned money or rather the lack of it. Cockayne wrote: 'McMeekan has been experiencing great difficulty with his financial position and representations have been made to me for further assistance. I approached the Dairy and Meat Boards and got £100 from the Dairy Board. McMeekan reckons he can get by, provided he can have his whole salary, as last year he was required to contribute £150 for his relief.' (McMeekan was still on the staff at Massey.)

Cockayne concludes: 'I hope you can arrange for Cooper to do the necessary lecturing at Massey without financial objections on the part of McMeekan.' This letter was clearly passed on to Cooper for comment and his pencilled remarks are added at the bottom of the original letter.

I am in sympathy with McMeekan's position. Mr Cockayne appears to recognise that McMeekan deserves assistance but it is presumptuous of him to ask a private individual to provide that assistance. I estimate that the work at Massey will require two days of departmental time and at least thirty hours of my own time, besides the inconvenience of spending one night per week away from home. It is impossible in view of my present salary to undertake the extra work without remuneration. At the same time I consider McMeekan's work so important to New Zealand that he should have assistance.

Cooper was keen to get back to teaching but he was not going to be exploited. Fainter hearts might well have got on with it and hoped that their diligence would eventually be rewarded. In any event, an arrangement was agreed, and Cooper began to plan relief lectures at Massey in McMeekan's absence.

There can be no doubt that the staff at Massey were thrilled to have Cooper back in the fold, if only on a part-time basis. It also seems that Cooper's job at DSIR was falling short of his expectations, despite the fact that he had recently been promoted. According to sister Thelma, 'Mac was not very happy in Wellington, writing speeches for Ministers and doing paperwork. He was too far removed from practical things and above all he wanted to work with students.'

Copies of correspondence with Professor Riddet in early June 1938, unsurprisingly therefore, indicate that Cooper intended to take the content of the lectures he was planning extremely seriously. Not for him was this a 'fill-in' job, but rather a golden opportunity to impress his former colleagues with his interest and enthusiasm for returning to teaching at the practical level.

An example from one of his letters to Riddet proves the point:

I have completed three of the special lectures for the Dairy Farmer's Course. I am enclosing two, which you have already seen for your consideration. I am not entirely happy with the lecture on calf feeding. The material that McMeekan had in his special lecture, I do not consider particularly good – but there has been a paucity of publication on calf feeding under New Zealand conditions and I have to rely very much on the principles of rationing and my own personal experience. If you would consider the statements I make carefully, I should be grateful, because the topic is one on which farmers may have decided views.

Apart from the lectures and their preparation, together with the travelling and staying away from home, there remained his ongoing work for DSIR. As he became more established he was seconded to more committees with the inevitable increase in administrative responsibility. Also imminent was their first child, due at the end of June, so life was, to say the least, extremely hectic. Rather than in Wellington, Hilary's confinement was to be in a private maternity hospital, over 300 km. away in Hastings, run by Mac's sister Madge. Being so far away from home, care was lavished upon Hilary and although the birth was straightforward, she didn't emerge from the hospital for well over a fortnight; an unheard of period by today's standards, when mothers are home within twenty-four hours of giving birth.

An announcement was duly recorded in the July edition of *Bleat* under the birth section:

We are pleased to be able to report here a successful effort on behalf of Mr and Mrs Mac Cooper to bolster up the waning birth rate of this country. The happy event occurred when there came into this world a bonny blue-eyed girl, Barbara Margaret, who should inherit a distinct enthusiasm for pigs and rugby. We feel that congratulations are due on this felicitous occasion.

Mac meanwhile, seemed not to appear to be the doting father. Engrossed in his work, he only managed occasional visits to Hastings, happy in the knowledge that mother and daughter were in very capable hands. They had no car and public transport was time consuming and expensive. Maybe this continued involvement in the rugby world was a limiting factor in his enthusiasm to travel over to Hastings. He had also joined the Wellington Club and within the space of a few months was elected captain, involving training and selection meetings in addition to playing every Saturday. Bill Brien, the Club Historian, has unearthed the match reports from the club records for 1938:

Possessing a highly efficient set of backs, the Senior Grade A1 team produced some brilliant displays, and it attained the very creditable position of third in the grade, after defeating in the last two matches the ultimate winners and the runners-up. The team was a distinct credit to the Club, and its success was due in no small degree to the work of the coach, Mr F.D. Kilby, and to the leadership of the captain, Mr M.M. Cooper.

The results for the following season were even better.

Warmest congratulations are due to the Senior A1 Team, not only on its notable achievement in winning the premiership for the first time since 1914, but on the very attractive manner in which it did so. Its style of game was decidedly popular, the sportsmanship and the conduct of its members most commendable, and the team will have a lasting place in the history of the Club. To its members, and in particular to Messrs. F. Kilby, coach, M.M. Cooper, captain and R.C. Veitch, vice-captain, we express appreciation of this distinguished performance.

Mac Cooper, the Wellington RFC captain, receiving the NZ Club Championship Cup in 1939.
(Wellington Evening Post)

Bill Brien also comments in correspondence concerning a photograph taken following the end of a crucial match in that championship.

The photo was taken in the Club Championship in 1939 just a few days before the outbreak of war. Wellington played their old rivals Marist Old Boys and it is generally regarded as one of the great club matches and the leadership of Mac was apparently outstanding. The crowds still standing on the bank applauding gives some indication. Mac is being chaired from the field by team-mate Adrian Thomas and opposing Marist skipper Clem Shannon. One cannot imagine in today's professional environment a winning captain being so highly honoured by the skipper of a team that has just been defeated. Of more significance is that three of the players in the photo were playing their last championship game at the famous Athletic Park as they paid the supreme sacrifice in World War Two.

Mac is also credited at this time with being one of the moving forces behind the formation of the Centurians Football Club. This initiative produced a side including several well known personalities who were nearing the end of their careers and together with up and coming youngsters, they played a few games at the end of each season as a means of encouraging interest in the districts. To be invited to play alongside these household names was, for ambitious

Wellington Rugby Club versus Marist Old Boys. NZ Club Championships 1939.

enthusiasts, a great honour. The concept was not far removed from the Barbarian Club in Britain, for whom Mac had played on a number of occasions.

In 1939 McMeekan returned to Massey following his work at Cambridge and a prolonged research tour in both the United States and South America. The July 1939 edition of *Bleat* welcomed McMeekan back but 'regretted that the College was unable to retain Mac Cooper as a member of staff.' It meant Mac's return to Wellington and the endless round of committee work and politics. However the seed had been sown and it was clear that as soon as an opening presented itself, Cooper would be back at Massey on a permanent basis.

With the advent of war, however, life was about to change. Mac Cooper had long since had an inclination towards military life, from trench warfare in the garden with his brothers during the First World War, through school as a cadet and at eighteen when he had joined the Hawkes Bay Territorial Regiment. He later transferred to the Wellington West Coast Regiment as a 2nd lieutenant. He therefore saw it as his duty to play a full part in the coming conflict in Europe. But it wasn't going to be as straightforward as he might have imagined.

Agriculture was the key industry in New Zealand and her future economic welfare depended on being able to produce food more efficiently and cheaply than anywhere else in the world. This may not have been true before the war

Mac had always had an inclination towards military life.

67

but it would undoubtedly be true after it. Those like Mac Cooper who already had reputations as up and coming prospects in agricultural education, were therefore to find that their value to the nation was considered to be better served by remaining within their existing professional spheres rather than embarking to a foreign battlefield.

His opportunity to return to Massey as a full time lecturer in Dairy Husbandry came sooner than he had anticipated. Early in 1940 the chair of Animal Husbandry at Lincoln in the South Island became vacant and McMeekan was duly appointed, leaving the door wide open for Cooper to follow, again in his footsteps.

So it was without regret that Cooper said goodbye to the institutionalised life in Wellington and moved his family up to Palmerston North, first to a house in Park Road, and later to Te Awe Awe Street. As if to celebrate the event, the family expanded to five when the twin girls, Diana and Cynthia, were born in September 1940. No incarceration in Madge's nursing home this time; Hilary was determined to have Mac within earshot so that he could shoulder his fair share of the domestic burden.

There were two major consequences for Massey as a result of the war. First, many of the able-bodied staff volunteered for territorial service and whilst still keeping their teaching duties going, a considerable amount of time and effort went into military training. In 1941 the army established a Staff Training College on the campus. This was clearly an initiative encouraged by the Principal, Professor Peren, who as a Brigadier in the Territorial Regiment commanded B Squadron, Manawatu Mounted Rifles (Mechanised). Fortunately by then they had graduated from horses to vehicles. As might be expected, Mac Cooper transferred to the Manawatu Rifles on his arrival at Massey and soon began helping to recruit new members into the squadron from the student body.

The second significant effect on the College was the inevitable decline in the number of students. By 1942 enrolments were down to 48 with only 12 full time and 11 diploma students in residence. In fact, it was the British Government who specifically encouraged the continued recruitment of students at Massey as they recognised that increased food production would become a major priority as the war progressed and the need for trained agriculturists would become of paramount importance. Student numbers in Britain were almost non-existent, as by then many of the Colleges had closed down for the duration of the war.

Teaching at Massey was, therefore, grinding to a halt and although the research activities on the farms kept Mac Cooper busy, he was becoming increasingly anxious to take a more active part in the war effort. In keeping with many family men, he wanted to do his bit and furthermore, he wanted to be seen to be doing it. Hilary on the other hand, with three small children and far from her people in England, agreed enthusiastically with the authorities that Mac should remain in New Zealand. Although a captain in the Manawatu Rifles, his requests to join the New Zealand Expeditionary Force to Europe met with little success and he had to be content with writing up Army Education Welfare Service Reports on dairy and pig farming in readiness for returning servicemen after the war.

He set about the task diligently and the publications were comprehensive, covering economics, breeds, production and management within each species. They were written in a practical, down to earth way so that they were of equal value to farmers and students alike. But as time wore on, he became frustrated with this desk-bound job and longed to get into the real action. In a letter to Dr Allen at Oxford, dated February 1943, Mac recalls his fallen contemporaries and expounds his views on things as he sees them.

Your Christmas letter has just arrived and it has aroused in me a queer mixture of nostalgia and shame. The reason for the former is obvious for one who enjoys such wonderful memories of Oxford. The latter comes through my being such a rotten correspondent. However, I will try and make amends.

I was very shocked to read of the death of several of my good Oxford friends, Trilby Freakes, Jim Mann and Dick Holdsworth. Freakes was a particular friend of mine and I owe a lot to his quiet sense of humour in those harrowing days for the rugger captain that drag onto the 'Varsity match. One can only hope that their deaths and the deaths of thousands like them will crystallise in the hearts of all of us a determination that ruthless slaughter won't prevail again.

But apart from the bad news your letter brought, there was a lot of good reading too. If my reaction is any criterion, then your lively news of wartime Oxford will be widely appreciated.

I am now at Massey and the place is very much in my blood. I judged myself a very fortunate man as I returned to it as second in command of the Dairy Husbandry Department, early in 1940. I had something over two years as a Civil Servant before I returned but though I don't regret that experience, I was very glad to have it behind me. While there, I did my best to try and improve the prospects of Rhodes Scholars returning to Government employ. But there is an immovable brick wall of fantastic regulations and rights of appeal by existing Civil Servants which prevents anyone returning to a position worth more than £315 per annum unless a certificate is produced by the head of a department that the position could not be held by anyone else already employed in that service at a lower salary. What hope has a Scholar who has done Modern Greats of getting in at a respectable salary to a worthwhile job under such conditions?

No, I cannot recommend our Civil Service to any returning Scholar.

But Massey is different. It is a beautifully situated College with a 'campus' that a lot of our American visitors envy. It is young, it is virile and it is working in a field which is the life of half the people in this country. Because my heart is pretty close to mother earth and because people, especially young men who are just starting on their careers, are my greatest interest, I will not feel my life is wasted if I can continue to do a worthwhile job at Massey.

But these are thoughts of the future and peace.

I had almost forgotten I was a soldiering man. I joined up in 1940 but after four months I was manpowered back to the College but returned to the army when the Japs started their sideshow in our direction. I spent last year as a subaltern in an armoured regiment but just recently I was posted to Divisional HQ as GSO in charge of army education. I suffer this fate through starting off a scheme of army education last winter with the help of a very enthusiastic young padre. Now it has been decided to make the scheme a national one and for my sins I have been given another pip (captain) to suffer under the very eyes of the 'brass hats'. Actually I am very thrilled with the appointment because this waiting around for things to happen has been a pretty deadly life for officers and soldiers alike. Now I am able to take arms against that tide of frustration that is the toughest enemy we have had to face so far.

A little family news. My wife, you may remember, was an Oxford girl and it was quite brave of her to come so far away from home to share my life. She has had her blue moments naturally but she has never regretted her choice. Her life has been a very full one because our family grew rather rapidly. Our eldest daughter who is rather better than an average specimen (and that is not just fatherly pride) had her wish for a baby brother fulfilled in the shape of identical twin sisters, who are now two. It is utterly impossible to get any help now and so my wife has had a pretty solid time. We husbands can dodge off to camp and leave behind our domestic worries, but with a wife it is different. As soon as the war is over she is going back home for a trip which I hope will be some reward for her courage and industry these last two years. I am afraid it will be no holiday for her for she will insist on returning with all the Cooper brood to display to adoring grandparents.

May I say again how much I enjoyed your letter. You spoke of the reddening copper beeches of Wadham. I bought a house here with a very sweet garden, largely because it had a young copper beech and because it had lilac to remind me of north Oxford in May.

My best wishes to you and Mrs Allen and if old Fred is still at Univ., tell him I will give him a box of cigars if he can give my name when I next go into the lodge.

In his reply, Allen sympathises with the problems of New Zealand Civil Service protectionism and waxes eloquent about the same problem elsewhere in the world.

I am very sorry to hear what you say about the prospects for Scholars in the civil service, not only from the point of view of the Scholars themselves, but also from that of New Zealand.

However I console myself with the thought that there are perhaps quite enough civil servants in the world already. The proportion in this country nowadays, under war conditions, is really fantastic; that cannot be helped I suppose in an emergency, but the anxious question is whether it is going to continue after the war, for bureaucracy is a strong vested interest.

I cannot think that it will make for great vitality in the nation if all our best educated young men are going to be mere officials, and I am somewhat concerned at present at the large number of our ablest young American Rhodes Scholars who are all gravitating to Washington. Some of them are doing valuable work but I am a little sceptical about the size and complexity of the new American bureaucracy.

At last in June 1943, Mac Cooper finally persuaded the College authorities and the army boffins that if he were to have any credibility as an effective lecturer to returning ex-servicemen after the war, he must share their experiences in active service. That did the trick. He had to relinquish his rank, but he got his way. Army records are extremely sparse with regard to locations of personnel during wartime. Letters home were similarly limited in factual content, as all were subject to censor in case they fell into enemy hands and gave away sensitive information. Furthermore, Mac, like many other war veterans, talked little afterwards about his experiences. Nevertheless the NZ Army 'Cooper file' reveals that he embarked from New Zealand bound for Egypt on 21 July 1943. This was part of the 10th re-enforcements of the 2nd NZ Expeditionary Force and he was to serve with the 22nd Motor Battalion as a lieutenant (temporary).

A brief history of the 2nd Division's campaign in the Middle East during this phase of the war is given in the *Encyclopaedia of New Zealand*. A volume of the *Official War History of New Zealand*, compiled by Jim Harrison (published in 1958) is devoted to the 2nd Division and this provides some indication of some of the activities which Mac Cooper must have experienced as he progressed up Italy.

The 2nd Division had been heavily involved in the desert war in North Africa under the command of Major-General B.C. Freyberg. By 1943, the men were seasoned veterans, having returned to base camp at Maadi, near

Cairo, after fighting at Tobruk, El Alamein and in Tunisia. A substantial number of the men in the Division went home on well-earned leave with reinforcements coming into replace them. Although this meant some dilution in experience, the Division had earned such a reputation that Churchill himself urged that it should be allowed to continue with the Allied advances into Europe.

The 22nd Motor Battalion had been taking shape for over a year and whilst to many this was a frustrating time of inactivity, the result was an outfit which was highly mobile, self-supporting with its own vehicles, and possessing strong firepower. The campaign in Italy proved to be one of the bitterest experienced by the Allies in the whole of the war. Indeed the conditions endured by many of the front line troops at Cassino were described by various commentators as horrific as those at Verdun and Paschendale in the First World War. To give some context to Mac Cooper's war, it is necessary to outline the background to the invasion of Italy in July 1943.

At the political level, the Americans were keen to focus all their attentions on the enemy by a massive invasion through France from Britain. Churchill, on the other hand, was adamant that following the victory over Rommel in North Africa, the Allies should continue their advance up Italy, which he famously described as the soft underbelly of Europe. He was also anxious to keep open the Mediterranean passage to the Suez Canal as well as ease the pressure on the Russian front by forcing Hitler to divert troops to try and halt the Allies' advance up Italy. In the event Churchill got his way, but the American Generals were unconvinced and felt that they had been 'led up the garden path'. Furthermore, they expected that at any time, they might be recalled to undertake what they saw to be the key priority, an invasion of mainland France.

A degree of mistrust, therefore, developed between the American and British commanders in the field, which was yet further complicated by the fact that the Allied forces included troops from France, India, New Zealand, Canada and Poland. Perhaps unsurprisingly therefore, the entire campaign seemed to suffer from a lack of clear strategic objective and at times woefully inadequate communications.

The Allies invaded Sicily on 10 July 1943. Whilst the initial landings held the advantage of surprise, there was much confusion and many examples of self-inflicted casualties as airborne troops sustained friendly fire and landed in the wrong place. Nevertheless a secure beachhead was established, Palermo was taken, and the Allied forces proceeded to advance across the island towards the Straits of Messina. It took, however, until 17 August to overcome the resistance and the enemy was able to achieve a largely successful withdrawal of over 100,000 men and much of their equipment onto the mainland. According to *Great Battles of World War II*, published by Marshall Cavendish (1995):

For the Allies, Sicily was a painful transition from a North African to a

European war, while the Germans conducted a most successful rearguard crowned by a brilliant evacuation. Between 11-17 August, they brought all their units and the bulk of their equipment out of Sicily. More ominous was the confidence Sicily had given the German High Command to make a fight for Italy rather than abandon the peninsula entirely. The war in the area had only just begun.

The enemy, as opposed to the Allies, therefore, clearly had a strategy. They were going to provide stiff resistance based on preparing a series of strong defensive lines utilising the rivers and natural topography of the Italian mainland. The most notorious of these was known as the Gustav line and stretched across the peninsula, south of Rome from Minturno in the west to San Vito on the Adriatic coast. The key defensive position in this line was the infamous heights of Monte Cassino.

By the end of July, and the imminent invasion of the mainland, the Italian people had had enough. Mussolini had been overthrown and had fled to the north, where he had been installed by Hitler as the Head of a Fascist State based at Salo on Lake Garda. The King of Italy and his Prime Minister were keen to seek an armistice with the Allies, in the hope that such a policy would achieve a measure of leniency in the terms of the approaching surrender. In fact they went further, by declaring war against Germany in early October and ordering their forces to change sides. This turn of events caused little problem with the Italian air force and navy in that they simply defected to Allied bases with their equipment. For their army, however, it created chaos. Most of the ground troops were actively engaged with the Germans, fighting on both sides of the peninsula to slow down the Allies who by now were advancing steadily north on both coastal plains. Behind them the Germans knew they had the relative security of the Gustav line to fall back to.

With the surrender, many of the Italian troops were unsure whether to lay down their arms or start fighting the Germans. Initially, many were encouraged to take the former course and they were allowed to return home to civilian life. The Germans, however, were furious with their former allies who had deserted them in their hour of need. As the war progressed their attitude hardened and the remaining Italian forces were given the option of either joining their ranks or being deported to forced labour camps back in Germany. In the event this tactic resulted in many Italian soldiers deserting to join the growing band of Partisans and the acceleration of a long and bitter campaign of sabotage. The Germans showed little mercy with either the Partisans or those who sympathised with them and as the last vestiges of civilised behaviour crumbled, reprisal executions and massacres of women and children became commonplace.

This was the background to the start of the war in Italy as Mac Cooper's battalion landed at Taranto in the heel of Italy, in late October 1943. The 22nd

Motor Battalion, as part of the 2nd NZ Infantry Division, was to progress the 250 miles to the front line in the east to join up with the British Eighth Army.

On the other side of the peninsula the American Fifth Army, having narrowly avoided being pushed back into the sea following their mainland invasion at Salerno, were slowly and painfully progressing towards the western end of the Gustav line, and Cassino.

In his book *Cassino – The Hollow Victory*, John Ellis describes the conditions which faced the Allied forces as they fought their way north.

> Italy is a country where much of the terrain offers an endless series of obstacles to an advancing force. Below the northern plains one simply finds one great mountain range, the only low ground being the narrow coastal strips and the river valleys which fan out from the central spine like spindly fishbones. Each of these river courses offered a superb defensive position, the Germans usually flooding the attacker's side of the river, building deep dug-outs and machine gun posts on the far bank and siting their artillery in the mountains behind. After their near shave at Salerno, the Allies spent the rest of 1943 banging their heads against a whole series of such positions.

Conditions were similar on both sides of the central Apennine mountain range and the New Zealanders arrived at the River Sangro after a non-hazardous, but frustrating journey with 4,500 vehicles struggling through muddy roads in appalling wet weather: a very different environment to that experienced in the North African desert. Mac Cooper's battalion was assigned the task of protecting the Division's left flank as it advanced to the river. In fact, this proved to be relatively straightforward as the enemy had decided to push back to its winter line on the far side of the river, so avoiding a major showdown. For the crossing, therefore, the Battalion was in reserve.

The main attack, involving the Eighth Army, came in the last few days of November, with tanks and infantry. It was as well that the resistance was slight as the heavy machinery made slow progress in the sodden ground each side of the river. Success was, however, achieved and the engineers managed to establish two bridgeheads across the river to facilitate the crossing for the bulk of the Forces. This was a good start to the campaign for the New Zealanders, despite the physical difficulties caused by the weather. It was the first time many of the reinforcements had come under fire and the apparent ease with which the operation unfolded gave them much needed confidence. Communications between the various units had also appeared to have worked well and as a result morale took a significant step in the right direction.

It must have been in this area that a letter from Mac Cooper to his sister Thelma was posted. Dated 4 January 1943 instead of 1944, the letter clearly intended to reassure those back home that life was not that bad and that he was not in immediate danger. He was also careful neither to divulge his whereabouts nor hint at the real conditions facing the front line troops.

Mrs J.F. Porter   471244 Lt. M.M. Cooper   date Jan. 4 '43
Whakarara   22 BN   2NZEF   MEF
Waipawa
New Zealand

My dearest Thelma and Joe,
Perhaps because the countryside, with its blanket of snow, six inches deep, reminds me rather of your home district, I have been thinking a fair amount about you lately. This snow is not nearly so bad as the BBC paints it, but it is bad enough, though we are having the redeeming factor of bright sunny days which make the countryside truly beautiful. We are now billeted in a little peasant village about five miles from the front line, which is a very good thing for I am no 'ero when it comes to these over-sized throw-downs being tossed about. Life these days on the whole is fairly uneventful and so I have no complaints. I am extremely fit and well and weigh over 14 stone, which is a tribute to the really wonderful rations we are getting. We certainly do not want for anything in respect of clothes or food. The one thing I need, but of course can't get is the presence of my beloved family. I have a lot of deferred happiness in store. Peasant life here is pretty primitive, as I have described in letters to Madge, which I presume you have seen. We had a very excellent Xmas, thanks to a great effort on behalf of our cooks, who produced a meal worthy of the Ritz Carlton. I hope this note finds you all fit and well. I hear about you from Margo, but I never despair of receiving a letter in your handwriting – or perhaps Sally's [a niece]. I hope you are having a good year on the farm and the fruits of your labour are beginning to ripen in a worthy crop. We have as one of our cooks, a leading citizen of Whakarera, one Allen Boden, a genuine little chap.
    My love and best wishes to you all for the New Year.

Mac.

Although Mac Cooper had managed to get into active service, his perceived value to the authorities back home would mean that he was still to be denied front line duty. Try as he might, the orders were that he was to stay in the background. Terry McLean, later to become Sir Terry and a well known sports journalist, was in the same Battalion and writes of Mac Cooper:

He was one of life's enthusiasts in whatever he took up. Haddon Donald, a surviving Commanding Officer of the time, remembers him as a capable and conscientious officer who throughout his tenure with the Battalion was 2nd in command of a company whose job it was to look after the care, clothing and victualling of the troops.

Taking rations, mail and other items forward was not a cushy job, especially at a place like Cassino, which had to be reached by route 6, including a bridge over the River Rapido. The Jerries had both the road and bridge accurately targeted and more than once we lost members of our rationing parties.

No doubt the 'miracles' worked by the cooks at Christmas, if they are to be believed, had something to do with the rations Mac and his colleagues were able to procure. Anecdotal evidence also seems to indicate that in his enthusiasm to supply the men up front, he managed to end up on the wrong side of the line on a number of occasions.

The 'feel good' factor following the crossing of the river Sangro before Christmas, however, did not last long. After making some progress into the Abruzzi Hills to the north of the river, the Division's advance was decisively stopped at Ortona, a large village with natural fortifications and doggedly defended by a paratroop garrison. Over four hundred New Zealanders were killed in the fighting early in the New Year before the Division was relieved in mid January. Worse was to come as they then set out across the Apennines towards the West Coast and the battle for Cassino.

St. Benedict founded the Monastery at Monte Cassino in AD 524. Perched on top of a precipitous hill overlooking the town and with a commanding view of the main road to Rome (route 66), it was the key point in the western part of the Gustav line and the Germans had orders to hold their ground at all costs. It took four assaults against impregnable defences and six months to breach the line. Many war historians describe the sequence of battles as the bitterest in the whole of the Second World War, with a total of over 200,000 men being either killed or wounded in the fighting. When the New Zealand 2nd Division arrived in February 1944 to relieve some of the Americans who had been beaten back by remorseless German fire power, they were horrified to see the condition of some of the survivors as they came out of the front line.

Ellis, in his book, quotes a New Zealander:

They were infantrymen returning to a bivouac area for rest and I knew from the divisional emblem on their sleeves that those men had been up in the mountains around Cassino. I thought I had never seen such tired faces. It was more than the stubble of a beard that told the story; it was the blank staring eyes. The men were so tired that it was like a living death. They had come from such a depth of weariness that I wondered if they would ever be able to quite make the return to the lives and thoughts that they had known.

The failure of the Americans to break through the line did not discourage their commander, General Clark, to repeat the process, believing, as Ellis records 'in the best traditions of the Somme and Paschendale, that just one more push would do the trick'. The New Zealanders, together with a number of Indian divisions, under General Freyberg, formed the New Zealand Corps and they were to be the ground forces to be used in this second attack.

Prior to this attack it was decided that the Monastery itself should be bombed, as it was believed that the enemy were using the building for artillery observation. This was a highly controversial decision at the time as there was no hard evidence that the Monastery was being used in this way and the Germans had stated it was their intention that this historic monument should be preserved. In the event some 600 tons of bombs were dropped as a softening up process, followed up by the infantry and armoured assault on the town and the Monastery.

With the benefit of hindsight, the plan was doomed from the start. Rather than eliminate the defences, the bombardment actually created more easily defended positions in the heaps of rubble within the Monastery, which the defenders occupied with some enthusiasm. It also made progress by tanks and armoured vehicles in the town virtually impossible as collapsed buildings and bomb craters made most of the roads impassable.

Ellis concludes this catalogue of events with the following opinion:

The New Zealand Corps failure was not a result of bad luck, not even of slapdash staff work that on at least two occasions had kept 7 Indian Brigade woefully misinformed, but a wilful failure at the highest level to take account of the terrible problems involved in mounting a concerted attack across such appalling terrain. Worse still, that these problems were still being grossly underestimated a full month later.

After the NZ Corps withdrew the weather further deteriorated so that there followed a so-called lull in the fighting, during which a new plan was drawn up. As Ellis points out, however, this description is misleading.

If the weather was an inconvenience to the senior commanders, it was an appalling hardship for the troops actually in the line. This is a point that should not be overlooked. The fighting soldier's lot in World War Two did not consist simply of sporadic flurries of combat interspersed with long reliefs or rests. It was an unremitting battle against nature itself, in which the so-called 'lulls' in the fighting were spent in holes in the ground, usually cold and wet and almost always within range of the enemy's machine guns, mortars and artillery.

The new plan of attack, scheduled for mid March, was again to be preceded by aerial bombardment with rapid follow up by ground troops, albeit from a different direction. But this too was to end in a failure to make the decisive breakthrough into the Liri Valley beyond and to open the way to Rome. The reasons were fundamentally the same. The bombardment whilst flattening what remained of the target failed to eliminate the enemy. The approaches to the town were limited either by exceedingly difficult terrain or by badly damaged and exposed roads. Once inside the ruins the defending troops fought furiously and progress for armoured support was virtually impossible because of the devastation.

Ellis quotes New Zealand memories.

The bombardment was apocalyptic. It was amazing anyone survived it – but survive it they did and they were enraged, even killing Red Cross people. Also the attackers were furious and any residual notions of chivalry and fair play soon evaporated and were irrelevant.

Said another New Zealander: Marching into the ruins brought out a kind of sadism in us. We were really enraged when the Jerries hit back – they had no right even to be alive. They were behaving like machines. Some bugger near to me was hit and he went screaming straight towards the Jerries, not in pain but in sheer fury, like some frantic animal, scrabbling through the ruins. Of course they got him.

As one anonymous New Zealander put it: 'The name that is most generally applied to the NZ Division is Kiwi. One version of the derivation of the name I have heard goes like this: like the bird we can't fly, we can't see and we are rapidly becoming extinct.'

This, the third offensive, ended on 24 March and two days later the New Zealand Corps was disbanded. Isolated detachments on the surrounding hillsides withdrew with great difficulty and all were bitterly disappointed that yet again, they had failed in their mission to break through the line. In fact it was not until May with the improved weather conditions that 11th Polish Corps successfully dislodged the Germans from Monte Cassino. However, by then, enemy strategy was to withdraw north to the next stronghold, known as the Gothic line, which stretched from La Spezia on the Mediterranean in the west to Rimini on the Adriatic coast. Rome eventually fell to the Allies in June 1943 but the cost had been high. Furthermore significant numbers of the enemy had escaped to fight on for another winter in the north.

Ellis, near the end of his comprehensive analysis of the fighting at Cassino, states:

The Allied fighting troops had been brought to the very extremes of physical and mental suffering. Icy cold, rain, sleet, snow, mud, hunger, fatigue, the crash of shells, severed arteries, exposed innards, exploding brains, and fear, gut wrenching, bowel moving fear – these had been the motifs of Cassino. Yet amidst this infernal ambience men had attained epic stature, keeping tight hold of themselves during their 'long days dying' and rising when bidden to trudge forward into fire up the slopes of yet another godless, barren hillside. And now all they had to look forward to was yet more of the same. Truly had Cassino been a hollow victory.

About this time Mac Cooper wrote to Dr Allen at Rhodes House in reply to his regular Christmas letter which he circulated to all old Scholars. Again it gives not the slightest indication of the conditions which prevailed on the battlefield.

As usual, your letter gave me tremendous pleasure, perhaps more than usual for I live in hopes that I will be seeing something of Oxford before the year is out. It arrived, too, at a time when all of us were feeling rather fed-up with inactivity forced on us by the deplorable weather that has been our lot these last three weeks. However, the weather has steadied somewhat and perhaps in a few days we may be moving on our way to Oxford via Rome, Milan and Berlin.

When last I wrote, I was GSO2 for Army Education in one of our Divisions at home. As Jap threat to New Zealand became less acute, I began to feel rather embarrassed by my lack of contribution to the conclusion of the war and I applied for a transfer over here in a combatant role. It meant dropping rank but that was a small consideration for apart from the immense personal satisfaction I get from being here, I am also getting valuable experience of a body of men with whom I will be working on my return. In fact, this last point was the deciding argument in the tussle I had with the College when I wanted to get away. The College will be dealing with large numbers of returned men in the rehabilitation programme and I will be the only senior member of staff who is himself a returned man.

Naturally it was very bad leaving my wife and family but my wife was equally keen with me that I should join the ranks of so many Oxford friends. Both of her sisters have had their husbands on active service from the earliest months of the war. One of them, Derrick Lorraine, who was rugger captain in 1934 when I got my blue, has been a prisoner of war since Dunkirk.

I know I have left my wife with a big responsibility of three young children (including our heavenly but devilish impish twins) but she has a fortunate capacity for such responsibility.

She can't help but have rather grim periods but both of us are thoroughgoing optimists and consider we will be well repaid by the joyful reunion that will be ours. Separation like this certainly gives one an appreciation of the essential values of life.

After Cassino and the fall of Rome, the Allies continued their way north. Mac Cooper's Motor Battalion, now back within the Divisional command, remained in a supporting role, clearing mines and booby traps as well as repairing roads and bridges. They were disappointed not to be able to employ their undoubted strength and firepower, but they were rewarded for their patience as they fought up towards the River Arno and Florence beyond. It was not until mid August that this ancient and beautiful city fell to the Allies after a long and bloody spell of attack and counter attack costing the Division almost another three hundred lives.

By Christmas 1944 the Division had taken Faenza and joined up with the advancing Allies in the east, so completing the line across the whole of northern Italy. This was as far as they would go until the following spring. The rest of the winter passed quietly enough as the men rested and retrained before making the final assault in Italy on the Gothic line.

Shortly before the battle for Faenza a radio commentary of a rugby match on 8 December was relayed back home to New Zealand. This was the final of the New Zealand Divisional Rugby Championship between 22nd Motor Battalion and 2nd Ammunition Company, the trophy for which was the Freyberg Cup. The match was held in the ruins of an impressive stadium which Mussolini had built in Forli, a few miles east of Faenza. The match was played in atrocious conditions but with great enthusiasm before about five thousand excited spectators. The commentator, Terry McLean, in his pre-match build up, mentions that there were a number of well-known players on parade – including Bob Scott and Wally Argus, both members of the Kiwi team that toured Britain after the war. McLean also referred to 'Mac Cooper, who lectures at Massey, was a Rhodes Scholar and played for both Oxford University and Scotland. He's a bit past it nowadays but everyone is delighted that he has turned out.' In fact he was thirty-four. After the game, which the Motor Battalion won narrowly by 4 points to nil, General Freyberg began presenting the cup to the Ammunition Company. Apparently he thought that they had deserved to win!

To have conducted a regimental rugby competition through such traumatic times says much about New Zealanders. To them, rugby is not merely a game; it is more an attitude to life. John Mulgan, the New Zealander who may have unwittingly been the cause of Mac Cooper gaining his Rhodes Scholarship to Oxford, had joined the British army and after a distinguished service record wrote about his life and philosophy during the war. In his book *Report on Experience* he throws some light on this attitude to life.

Rugby football was the best of all our youthful pleasures: it was religion and desire and fulfilment all in one. Most New Zealanders can look back on some game which they played to win and whose issues seemed to them then a good deal more important than a lot that happened since.

This phenomenon is greatly deprecated by a lot of thinkers who feel that an exaggerated attention to games gives the young a wrong sense of values. This may well be true and if it is true, the majority of New Zealanders have a wrong sense of values for the rest of their lives. But to be frank, and since we live in a hard world, and one that has certainly, not in my time got any softer, I found in wartime that there was considerable virtue in men who had played games like professionals to win and not, like public schoolboys and amateurs, for exercise.

So that perhaps it would be more correct to say that the virtues and values of New Zealanders were not so much wrong as primitive, and to this extent useful in the current collapse of civilisation. New Zealanders, when they went to war, found it easier to get down to the moral plane of the German soldier, and were capable of thinking a ruse or two ahead in the game of total war. Englishmen spent some time and casualties in finding war ungentlemanly before they tossed the rules overboard and moved in on the same basis. I don't know that cunning and professionalism of my fellow countrymen is to be commended on abstract grounds, but these are comfortable qualities to have about in wartime. Oddly enough I don't think these things affected their natural kindliness nor the kind of ethics that they expect from people in private life. It was only that they looked on war as a game, and a game to New Zealanders is something that they play to win, against the other side and the referee if necessary.

The spring of 1945 saw the final assault begin. Although the end was in sight, the fighting continued to be ferocious with the enemy desperately defending their final positions. By April it was only a matter of time until Venice fell, and with it on 2 May came the official surrender of the German forces in Italy.

The total losses sustained by the NZ 2nd Division throughout the Italian campaign amounted to 6,581 dead, and 16,237 wounded. In the opinion of the authors of the *Encyclopaedia of New Zealand*: 'For the operations carried out, these totals are not high.' Although the war in Northern Europe was to continue for some months, the Division took no further active role in the fighting and until demobilisation commenced in the autumn, the troops remained in the area as an occupying force keeping the peace.

Mac Cooper's war, however, had been over for some time. He had been promoted to Major and although the Motor Battalion had been disbanded and the men posted back to their provincial companies, he was asked by Divisional Headquarters to travel to Austria and help establish a rugby training camp in Klagenfort. Mac was second in command to a Major V.C. Butler, an All Black full back, of the 21st Battalion and they proceeded to locate the best squad possible throughout the ranks of the 2nd NZEF in Europe. The squad of twenty-five players then travelled to Britain with Cooper for the final trials, to select the Kiwi side to tour France and the UK once hostilities finally ceased. Mac did not stay with the touring party for the whole of the time and after visiting his in-laws, by now back on Boars Hill, and former friends at Oxford, he returned to Italy at the end of November.

The rugby tour was a huge success and according to national press reports at the time, 'it resuscitated international rugby after the war and the players set new standards of innovation and entertainment'. The Kiwis competed in 33 matches, winning 29 of them, drawing 2 and losing 2. They beat full international sides in England, Wales and France, although they narrowly lost to Scotland, perhaps to Mac Cooper's secret satisfaction. Significantly, the players were not All Blacks; they were soldiers selected after service in North Africa and Italy. Some had even been prisoners of war. So, once again, rugby football became the symbol which New Zealanders adopted as a sign of returning normality.

Reference to his army file indicates that Mac Cooper left Italy, homeward bound, on 10 January 1946. On the face of it, this seems rather a long time after the war had ended and perhaps rather surprising, in view of his obvious desire to be reunited as soon as possible with his wife and family. There is no record of his activities during this six-week period, although it is likely that he was simply awaiting the availability of homeward bound transport. One can only speculate as to his thoughts about the two and a half years of war and deprivation.

As an enthusiastic volunteer and with a determination to play his full part, he probably felt 'short changed' when he found that he was to be confined to a supporting role, however important in supplying the front line troops. When, however, he saw for himself the life expectancy and appalling conditions which those men had to endure, it would be surprising if he did not feel a sense of relief that his original request had been turned down. Also, in keeping with his comrades, he would look back with profound disappointment, having endured so much sacrifice, that the Division had failed to achieve its objectives at Cassino.

Ellis in his book sums up his opinion of their possible reactions:

In the event, the New Zealanders were unsuccessful at Monte Cassino. Maybe it was bad planning or unassailable defences, as by then the German positions had not been totally destroyed. Or maybe it was true that the German resistance, bordering on the fanatic, was too much for them. This would not be an easily accepted result – they were after all, playing the game to win and they did not take lightly to being seen as unsuccessful.

But whatever his memories, Mac Cooper would return a changed man to a country and to a family who could never hope to understand what he had witnessed. Little wonder, he and hundreds of thousands of others who survived these and other battlefields, preferred not to talk of their experiences. Many felt guilty that they had survived while so many of their comrades had perished. It would be difficult enough to try and simply lead a normal life and the priority would be to think positively about the future rather than dwell on the horrors of the past. Maybe these were some of the reasons he never collected his four war service medals. No one knew they existed until they were revealed in his army file as part of the research for this book.

6

THE CROSS-ROADS

On the long voyage home, Mac Cooper must have thought deeply about his future employment. At the age of thirty-five, he had already achieved much, but he knew he was approaching a crossroads. He had established an enviable reputation at Massey before he went overseas and he had a worthwhile job to go back to, which was more than could be said of many of his compatriots. In addition there was the prospect of a chair in the Animal Husbandry Department, provided the funds could be secured to establish it. He had also achieved his ambition to take part in active service during the war and he had done so with some distinction, even if his role had been further from the firing line than he originally intended. At least he had come through unscathed and he would have earned the respect of the men who would be coming to Massey as post-war students. But there was another major consideration, which would have been very much in his mind. Hilary had not been back to England for almost

Hilary and the Three Whirlwinds, Barbara, Diana and Cynthia, New Zealand, circa 1945.

nine years and her parents had not even seen their grandchildren. He himself had missed forever some of those early years in his children's upbringing. He knew that they might not recognise him when he walked through the door.

For Hilary, adapting to life in New Zealand before the war had not been easy. Whilst she had been warmly welcomed by Mac's family, she, like so many other female English immigrants, found the change in lifestyle and culture difficult to comprehend. At Oxford she had been at the centre of a social round of dances and parties. She was to marry one of the rugby superstars and although her family had, at the time, been at low ebb financially, they were nevertheless from the top drawer of Oxford society. In Wellington and later in Palmerston North, she was a housewife with small children and with a husband who was immersed in his job during the week and who played rugby at weekends. Social life was very limited within a traditional and conservative country, where the place of the wife and mother was decidedly in the home. Those who dared to challenge the status quo caused much raising of eyebrows.

No copies of letters between Mac and his wife during the war years have survived, but it appears evident that they had discussed what they planned to do after the war was over. Hilary was intent on coming home to England.

In the Rhodes House file there is a letter from Hilary's father to Dr Allen, which is indicative of her influence before Mac arrived with the rugby players.

Mathews & Son (Insurance Brokers) Ltd.
6 St. Aldate's
Oxford   17 August 1945.

Dear Sir,

I am taking the liberty of writing to you with regard to my son in law, Mac Cooper, who is at present in the army. He was in Italy but I assume now he is in Austria.

I understand through letters from my daughter that they are desirous of coming back to England as soon as practicable, also that Mac is applying for a post at Aberdeen University, presumably in connection with agriculture.

I understand that Nuffield College is advertising for applications in respect of two posts, one at £600 and the other at £1200 per year, but I have no details with regard to the type of work involved.

I assume now that agricultural research will be expanded in this country and as I understand you have a good opinion of Mac, I am sure that if you hear of any position which you consider would suit him, you would think of him.

Yours truly,
Wilfrid Mathews.

Dr Allen replied that he did indeed have a high regard for Mac, and although he was unsure of the sort of job he might be looking for, he would help in any way he could. He assumed that when Mac arrived in England he would make contact and they could talk about it in depth; however, with the return of so many ex-servicemen, the competition would be very strong.

Perhaps when he had been in England, Mac did visit Dr Allen, but there is no record of such a visit. He certainly visited Hilary's relations and no doubt many other friends and contacts in the country. Perhaps it was then that he realised how difficult it was going to be to get a job he really wanted to do. Inevitably there would be preference given to returning British ex-servicemen and furthermore, it would seem unreasonable to deny someone else an opportunity when he already had a guaranteed challenge back home.

Whatever lay in store, he knew that he was going to have to devote a great deal more of his time to his wife and children than hitherto. Maybe with his return and the re-uniting of his family, plus the promise of an early trip home, he could persuade Hilary that their long-term future still lay in New Zealand. But as he neared the end of the voyage, maybe also deep down, he knew that they might all be making the return journey before too long

The returning troop carrier docked at Wellington on 10 February 1946. Anticipating a vast throng of people and general chaos surrounding the arrival of hundreds of returning ex-servicemen, Professor Riddet kindly suggested that he accompany Hilary to the quayside. It is as well that he did, for the crowds were overwhelming and had he not luckily located Mac quickly, frustrating delays might have marred their reunion. Riddet, having fulfilled his role, slipped away, leaving the two of them alone. One can imagine the welcome he received on seeing his young daughters and the rest of the family in the Hastings area. Equally his friends and colleagues at Massey were delighted to see him back safe and sound.

After the celebrations, it was back to work as Head of the Dairy Husbandry Department, a life which had the same sort of commitment to teaching, dairy research on the College farms, and extra-mural activities with students, as had been the case before he left for the war. This time, however, there was a major difference. Instead of declining numbers, there was rapid expansion, as mature students, intent on getting into farming, swelled the ranks. There was also a staff shortage with a lack of qualified people available to fill gaps, let alone cater for the increased numbers.

True to form, Cooper was in the thick of it and also found time to coach the College XV which, according to *Bleat* 'won for itself and its coach, many laurels in club and inter-Varsity football'. The student magazine goes on to say: 'it was in the 1946 season that one Saturday saw the XV one forward short during the game because of injury. The score was 6-6 when Coach Cooper took the field. He immediately added new life to the forwards and inspired the whole team to greater efforts. Mac soon proved to have lost none of his ability, when, with a kick from the halfway line, he finally added the winning points.'

But as the novelty of homecoming wore off, tensions began to develop in the household. Mac, in keeping with many ex-servicemen, found rehabilitation very difficult. He suddenly found himself in the midst of a family of three demanding children. He had to become an instant father to young girls whom he hardly knew and who did not know him. This time he couldn't just walk away and immerse himself in his work and rugby. In addition, Hilary had been alone for almost three years in New Zealand. She had not been back to England since she had married, almost nine years previously. She wanted to return and was determined to persuade Mac that this was in all of their best interests.

In August 1946, a cousin of Hilary's, Marygold Rix-Miller, arrived in New Zealand. She had married a New Zealander who had served as a bomber pilot in the RAF with great distinction and he was coming home to help run the family farm in the Waikato. Like Hilary, Marygold had great difficulty acclimatising to her new life and she records her experiences in a book entitled *Trophy of War*.

Inside the flyleaf of the copy she gave to Hilary, Marygold has written in a spidery hand: 'To my dear relative and friend Hilary, who understands.' In the book she describes the fraught relationship that she has with her husband's mother, with whom they and their young son were living until they could build their own house. Unlike Hilary, Marygold was in a country location, but her observations and her letters home to her mother throw some light on the different values which prevailed at the time.

I was beginning to realise there was a lot of local resentment against the English which I did not expect. Here and there small, slightly insulting remarks were passed which I did my best to ignore. I was amazed to find that English women were judged to be inefficient housekeepers, and every mistake I made was pointed out as proof of this.

I haven't met anyone who cares much about beauty here. One Englishman said, 'You've had it unless you can make a sponge cake and bottle fruit.' They certainly weigh you up on those lines. However they are so good hearted I don't want to feel critical.

I see the fare to Britain is now £101. Oh dear!

Another reference, on a similar theme:

I made so many stupid mistakes. I realised I was 'on probation' and I wanted so much to do well and fit in. There were very definite rules to learn in this local society I found myself in. I had to stumble around trying to get them right. I felt stupid when I obviously said and did the wrong thing. There would be an icy pause in the conversation – an intake of breath. I soon realised when I didn't conform to local standards and customs. The people made it clear I was the one who was wrong. They accepted no other way than their own.

In order to escape the claustrophobia, Marygold took her family to stay with the Coopers for a few days in Palmerston in October and from Te Awe Awe Street writes to her mother back in Oxford:

As you can see, I have come down to be with Hilary. She is just the same. Thank heaven she hasn't been influenced by New Zealand suburbia. I can't describe what happened when the two rebels got together. TALK!!! It was so wonderful to talk to someone who knew. We see things the same way and have just the same criticisms, so we tore New Zealand apart. Then having got that off our chests, we talked about all the things we liked and put it all together again . . .

Following their visit Hilary also wrote to Marygold's mother. Part of her letter gives cause for optimism.

Though it is nine years since I saw Marygold, I felt there is no gap. I think having talked to me it opened a safety valve for her. We just talked and talked. Mac and Hugh in the background making snide remarks about the English! I realise only too well the trauma that Marygold is going through and it's not easy, but she will come through with flying colours. New Zealand women are apt to put domesticity so far above everything else. It is going to be an enormous help to both of us to have each other. Marygold has done me more good than anything imagined.

It must be true that life in any close-knit rural community would have been more restricting than in a town. However these little cameos illustrate the stresses and strains which perhaps their New Zealand men-folk would, in all probability, dismiss as over sensitivity. In so doing, they no doubt made matters worse.

As the year at Massey wore on it became increasingly obvious that the hoped for Chair of Dairy Husbandry was unlikely to materialise in the foreseeable future. In fact applications for three Chairs had been put forward by the Governing Body and none of them had met with success. So apart from the tribulations of life at home, Mac Cooper began to wonder whether it really was time to make a move. Had the return to the predictability of College life, however enjoyable, begun to pall? Was there not a much greater challenge out there in a world trying desperately to come to terms with food shortages caused by six years of conflict? He was getting increasingly restless.

Another of John Mulgan's acute observations about New Zealanders which appears in his *Report on Experience* seems relevant at this time.

New Zealanders are a young people, but already with a place in history. They are often wanderers and restless and unhappy men. They come from the most beautiful country in the world, but it is a small country and very remote. After a while this isolation oppresses them and they go abroad. They roam the world looking not for adventure but for satisfaction.

New Zealanders are all the time standing on the edge of the seas. They spend their lives wanting to set out across the wide oceans that surround them in order to find the rest of the world.

One way or another, those who are going and those who are staying have all the time within them, this sad, inner conflict and frustration.

Reference to the Rhodes file shows that Mac applied for a position with the Food and Agriculture Organisation of the United Nations in Washington. A letter from a Mr McDougal, dated 27 November 1946, asks Dr Allen in Oxford for a reference. As might be expected, his reply is fulsome in praise with regard to character reference, but unhelpful in the context of agricultural matters. How many other applications he sent off and to whom they were sent is not known, but it seems probable that at around the time of Marygold's visit in October 1946, the Coopers had decided to leave Massey and New Zealand.

The background to Mac Cooper's appointment as Professor of Agriculture at Wye College, University of London, is somewhat vague. Most likely, the main players were Professor Scott-Watson, Head of the Rural Economy Department at Oxford, and a man called Dunstan Skilbeck, who had been a Demonstrator at Oxford in the Agricultural Economics Institute and who had recently been appointed Principal of the newly opened College following wartime closure. It may well be that Dr Allen at Rhodes House had also heard of the vacancy and notified Mac Cooper.

The news of the appointment broke in the New Zealand press early in the New Year of 1947. One report in the New Zealand *Listener* was aimed at a wider audience than the agricultural press.

In this country, Rugby is a much more familiar path to fame than learning, just as cricket or soccer is in Britain. To be learned in Animal Production, as M.M. Cooper is, to be head of the dairy husbandry department at Massey Agricultural College, which specialises in an industry worth so many millions to us, is not as spectacular or exciting as dancing behind an All Black scrum or flying down the touchline; though butter and cheese help to pay the wages (off the field of course) of these stars. A good many New Zealanders, one suspects, know little about Animal Husbandry. They might even be hard put to tell a Lincoln from a Merino, or a Hereford from an Ayrshire, and though they are fond of bacon, the proper care of pigs (who seem to need as much attention as babies) may be quite beyond their ken.

If M.M. Cooper had his way this would be changed. His main interest is in agricultural education, for which, indeed, he has a passion. He thinks all New Zealanders should know something about farming, including the farmer's mode of life, and he would have this life taught as a cultural subject in our University Colleges.

It was certainly true that Cooper's departure was used as a major lever to persuade the Department of Agriculture to grant permission for the previously requested Chairs at Massey. The one in Animal Husbandry which would, in all probability, have gone to Mac, went to Ian Campbell, who had been a former student at Massey, following which he had completed his post graduate work in America.

The reaction in England to Cooper's appointment was more muted. He was not known in agricultural academic circles and there were questions as to whether he had sufficient experience of British conditions to be able to handle the job. Rugby prowess was certainly not a qualification that cut much ice in post war Britain.

Some might have even suspected that the job had been fixed as he did not appear to have attended a formal interview and his credentials could only have been supported by Oxford academics, who hardly carried much weight in high places. One wonders whether Lord Bledisloe had had a hand in matters. He had been back in England for over ten years since he had been Governor of New Zealand and now led a very active life in areas related to farming and the countryside. He was an Honorary Fellow of University College Oxford and had Cooper's sponsors wanted high profile support for their man, Bledisloe, who had chaired their candidate's Rhodes Selection Panel, was an obvious choice. Interestingly also in 1946, Bledisloe was President of the Royal Agricultural Society of England and early the following year he visited Australia and New Zealand on a goodwill mission for the Society. It would be surprising in view of subsequent events if he had not talked to Cooper about the possibilities at Wye.

But it is likely that Skilbeck was the key player. He knew Cooper when he was at Oxford and as Principal of the new College he also knew what he wanted and he had the contacts and skill to get his way. He had an extremely forceful personality and decided views on how Wye was to function when it re-opened. High on his list of priorities was that it should be run on Oxford College lines with as many of its values and traditions as practicable being integrated into student life. His understanding of what might be required academically from the incoming Professor of Agriculture would have been superficial. His experience had been limited to the rather restricted and rarefied air of the Rural Economics Department at Oxford, a discipline that at this time was hardly taken seriously by agricultural academics. However, Skilbeck would not be concerned with such matters of detail, nor was he the man to rely on other's judgement as to the suitability for this post, academic or otherwise. What he did know was that Cooper was an Animal Production man, he had strength of character coupled with a sound reputation and above all he was an enthusiastic supporter of Oxford, its traditions and all it stood for. That was sufficient.

Whatever did go on behind the scenes, the London University Appointments Board eventually approved the recommendation and Mac Cooper got the break he had been hoping for. Not only did it open up new horizons in his career, it brought with it promotion to professorial status and most of all the opportunity to take Hilary and the children back to England. He jumped at the chance.

As might be expected the news was greeted with a mixture of congratulations and sadness at Massey by staff and students alike. The *Bleat* devoted a three-page eulogy to M.M.C. in their July 1947 edition.

M.M.C.
(By KSS and WGS)
Perhaps unique in the history of Palmerston North was the scene

witnessed on the afternoon of 10 June 1947. At one end of the Railway Station a band dressed in blue was playing while close by stood a group of young men wearing the characteristic tweed hats: for these were Massey boys farewelling their beloved Mac Cooper who, after escorting his wife and three children to their seats on the train, was carried shoulder high down the platform. Amidst handshakes and Hakas, cheers and confetti, the whistle blew, the train was moving – he was gone.

After a review of his many achievements both at Massey and during the war, the article concludes:

Wye College, Kent, where Mac has been appointed Professor of Agriculture, is somewhat similar to Massey in that research has been carried out coincidental with the training of students for the management of farms and professional positions in Agriculture and Horticulture. From now on, however, the College with a roll of over one hundred students is concentrating on degree courses only. Mac feels that, in his new job, he will be able to help New Zealand students overseas and be an ambassador interpreting the New Zealand point of view to English agriculture.

And so it was that on an afternoon in June, a spontaneous farewell was accorded – as only students can farewell – a man who is still a student at heart and who had grown up with Massey. A man who once said, 'We can't all of us be Professors of Agriculture, there have to be some to distribute the farm yard manure' – leaving us to take up a Professorship in England.

In his departure, New Zealand has lost for a time at least, a distinguished son and a gallant gentleman.

When Mac Cooper arrived at Massey in 1930, it had been a fledgling College with no history and no tradition. Wye was very different, at least in this respect. Stewart Richards has written a splendid centenary history of *Wye College and its world* which describes in some detail the foundation of the original College of St. Martin and St. Gregory by John Kempe in 1447. It goes on to record the major changes over the centuries from Grammar School to Agricultural College in the eighteen nineties.

Farming in Britain in the late nineteenth century (about the time James Cooper, Mac's grandfather, emigrated to New Zealand) was in a parlous state. A policy of free trade had encouraged increasing grain imports from North America and with the advent of refrigerated shipping, the southern hemisphere countries were beginning to supply substantial quantities of livestock products, thus undermining domestic prices. Such a policy had some appeal in that it helped to keep consumer prices down and at the same time it strengthened ties with the developing Empire. Predictably there was little public sympathy for the home producer and with a declining political influence, unlike some of the other European countries, the Government of the day provided no financial support for the farming industry. There was, however, a growing acceptance that the industry needed to embrace the new scientific discoveries which were being published, in the fields of both improved breeding methods and plant physiology. If farmers were to advance their prospects without Government support, then they must increase both their efficiency and their productivity. This meant that research results had to be put across in a practical and understandable way. How best to do this was the subject of much debate then and, in varying degrees, has remained a contentious issue ever since. On the one side were the advocates of an academic approach linked directly with those engaged in research. Through the higher education process they would ensure that the products of such a system were thoroughly trained in the science of agriculture and well able to influence those who worked directly in the industry, thus leading to improved practice on the farm. The farmers themselves were less than convinced that the academics understood their problems. They wanted a more immediate and practical outcome: students who were more proficient at the basic skills of ploughing and milking.

In 1893, the South East College of Agriculture was formally established at Wye and opened the following year under its first Principal, Daniel Hall. He was a distinguished academic from Oxford and is reputed to have welcomed

The main entrance: Wye College, University of London.

his first students with the words: 'We are gentlemen here, I take it, and your sense of what is fitting and upright will serve better than all the regulations I can lay down. Consider the College your home and leave the public houses alone.'

Hall had high standards and had recruited an impressive group of staff with which to establish a College based on Oxford principles. There existed a 250 acre farm which was easily accessible to the students, although initially, the running of it remained the responsibility of the County Council: further evidence that their farming representatives had little confidence in the academics to do other than teach.

By this time a number of British universities had established agricultural departments and Hall was anxious that his college should have formal affiliation at University level. Being sixty miles from London and in the middle of the countryside, it was little short of a miracle that he managed to convince the authorities that the South East College of Agriculture ought to be within the London University Faculty of Science. No doubt much of this achievement was due to Hall's own personal reputation, together with his staff's impressive track record in the previous four years. At a stroke it placed the College on a more challenging path, it opened up the potential in future to offer a London BSc degree and at the same time, Hall hoped to widen the vision away from some of the more parochial and constricting influences of local politics. Perhaps even more important, it would eventually ensure that Wye would be able to compete on equal terms for the best students throughout the country, so vital if it was to become the centre of excellence envisaged by Hall and his academic colleagues.

The next fifty years covered a period of tremendous change in British agriculture and in education. Wye was fortunate in having a succession of able Principals who built on the foundations laid by Hall and his pioneers. Unlike many of its competitors, the College managed to create for itself an identity which appealed to an influential sector of the market. It was a self-contained entity with a long tradition and buildings bearing close resemblance to the cloistered colleges of Oxbridge and the older public schools. A wide range of sports, societies and activities evolved over the years so that the students had all the resources, not only to gain a sound agricultural education, but also to hugely enjoy the experience. Although relations with the local farming community varied during this period, to say the least, Wye had justifiably earned for itself some credibility in running efficient farm enterprises. Furthermore, Wye graduates were also doing well both in this country and overseas, so the most important aspect of all, that of potential employment for those not destined to return to the family farm, became a reality.

This period of course included the two world wars and the recession between them. The policy of free trade pursued so vigorously in Victorian times came back to haunt the politicians when Britain had to rely so heavily on imports in war time. There followed a bewildering sequence of economic circumstances for farmers as agricultural policy swung from one extreme to another. Following the Government decision to provide financial support and to encourage home food production during the First War, farmers at last felt that they could make a real contribution. A degree of confidence prevailed, stimulating investment and innovation. But by 1920 the tables had turned, cheap imports poured in again and Government support evaporated. Little wonder that farmers felt betrayed with many of them facing bankruptcy after generations on the land.

For the next decade British farming was at extremely low ebb. The world wide recession simply added to the problem with cereal prices falling to lower levels than thirty years previously, resulting in a significant swing to livestock farming, permanent pasture and low cost production, simply to survive. By the early thirties, the free trade policy was finally abandoned and a new system of tariffs, quotas on selected imports and new marketing initiatives introduced. One of these was the Milk Marketing Board, set up as a statutory co-operative, to ensure that all dairy farmers received a fair price for their milk, rather than continue to depend on the whim of the local buyer.

At the outbreak of the Second World War, the academic activities at Wye were suspended for male students and arrangements made for the training of 180 members of the Women's Land Army. By the end of 1940, the College was requisitioned by the military authorities and only the work of the Research and Advisory Department, together with the farms, continued.

If the role of agriculture had had some importance in meeting the country's requirement for food in the First World War, it was relatively minor compared to the needs occasioned by the Second. At the time only about 35 per cent of the nation's food was produced at home and the enemy was well aware of the importance of cutting off supplies through an intensive submarine blockade of shipping in the Atlantic and Western Approaches. The predictable outcome was massive state intervention with the appointment of County War Agricultural Committees who had the powers to order farmers to cultivate permanent pasture, reclaim derelict land, and even force the eviction of recalcitrant tenants. These Committees had been used to some extent in the First War, so a great deal of valuable experience was already to hand, allowing them to be up and running quickly. The Committee members were a mixture of successful farmers, advisers and people involved in education. They were not, therefore, hidebound by bureaucracy. They spoke a language farmers could understand and in any case most farmers wanted to play their full part in the war effort.

Apart from the encouragement to use more subsidised artificial fertiliser, the biggest impact on increasing productivity came from the more widespread use of tractor power. The bigger farms, which were already mechanised, worked round the clock to help their neighbours who still relied on horsepower, so spreading the benefit over more acres. Some new tractors were manufactured as well as imported from America and as the power became available, so with it came the opportunity to develop new machines to improve cultivation, harvesting and grass conservation. Not only did these innovations increase productivity; they did so with a reduced labour demand. Apart from legislative encouragement, Government also supported cereal prices at the expense of livestock. As grain feeding to animals, apart from dairy cows, became uneconomic and millions of acres came out of permanent grassland, numbers of beef cattle, sheep and pigs declined, resulting in a sharp swing away from the 'dog and stick' farming of the previous twenty years. Although the national demand for food was much reduced by rationing, the outcome, in terms of substantially increased self-sufficiency, was achieved.

But if this policy of expanded food production in Britain was not to be reversed when the war ended, a more convincing and longer term strategy was needed to convince farmers that there would not be a repeat of the betrayal of the 1920s. This came in the form of the 1947 Agriculture Act which proposed that the main agricultural prices would be automatically underpinned by deficiency payments, funded from taxation, when owing to fluctuating supply and demand, market prices fell below those which had been guaranteed. There would also be an Annual Price Review by both Government and representatives of the National Farmers Union, so that a formula could be agreed to ensure farmers were able to make an adequate living. Furthermore, the Act introduced the concept of the Agricultural Wages Board, representing farm workers' interests. At last, there seemed to be a determination to put agriculture back on a sound and permanent footing, so that farmers might have the confidence to think long term. Some even started to plough some of their profits back into their enterprises, rather than keep the money in the bank ready to offset the next crash following another change in Government policy.

Wye College, no longer the South East Agricultural College, but now formally part of the University of London, re-opened in January 1946. Most of the students were ex-servicemen and for the first time women were to be admitted. Stewart Richards in his history of Wye refers to the 'great triumvirate' of Principal Dunstan Skilbeck, Louis Wain and Charles Garland as the key players in this resurrection of the College following the war years. He describes Skilbeck at the age of forty-one, as 'an establishment figure of impeccably upper-class credentials'.

> He had been educated at University College School, London and St. John's College, Oxford, where he studied rural economy during the period of J.A. Scott-Watson's Professorship. For three years he had then worked at the Agricultural Economics Research Institute before being appointed University Demonstrator in the School of Rural Economy, lecturer and tutor at St. John's and director of its farm. More recently he had served as Wing Commander in the RAF and Assistant Director of Food

Production at the Middle East Supply Centre. A charismatic personality and born leader, he spoke warmly of friendship but cared little if he made enemies.

Richards throws more light on this complex character when he quotes from the 1957/58 edition of the *Cardinal,* the College Union Society Journal.

Skilbeck's personal devotion to the estate and gardens, and his commitment to the highest standards of maintenance, became something of a legend in his own day. Nowhere more naturally than here could he indulge his fantasies as a feudal lord. The famous early morning rural rides ('the outside of a horse is good for the inside of a man!') would strike trembling obedience in the heart of many a cap-doffing labourer as he fled to open the next five-barred gate at his lordship's imperious command. Such impromptu tours of inspection must surely, he felt, be his secret weapon. Without them – that is, after his retirement – it would be quite impossible to guarantee the footprint of the master in every corner of the estate.

Peter Wilson, a student at this time, and later to become a prominent agriculturist and academic in Edinburgh, relates an incident involving his future wife. 'Bunny was, before I married her, an advisory entomologist in the National Agricultural Advisory Service at the Wye Centre. As such, she from time to time, visited farms in order to take soil samples for eel-worm counting, and on one occasion she was making holes with her sampling kit when up galloped Dunstan on his horse. "What the hell do you think you're doing?" he demanded. (Never a tactful man Dunstan!) "Taking eel-worm samples at the request of the farm manager," she replied. "Well, see you fill in all those little holes or you may cause my horse to stumble," growled Dunstan.'

The other two members of this so-called triumvirate were very different characters. Wain is described as 'a no-nonsense northerner'. He, unlike Skilbeck, was a 'grammar school and red-brick New-Man, whose limitless energy and enthusiasm for his subject left no one in any doubt that he was determined to get on'. Wain had three degrees and now, as head of the chemistry department, was set to make a very significant impact both at Wye and beyond, in his classical work on the control of plant growth and in the selective destruction of weeds. These scientific breakthroughs were to result in attracting both external funding and worldwide recognition for Wain and his team.

Finally, Charles Garland, as the combined 'Secretary, Registrar, Bursar, Appointments Officer, occasional Warden, Maintenance man, and even relief stoker', made up this formidable trio.

Richards describes him as the archetypal 'everyman'.

A personable individual whose charm and gentle leg pulling made him universally approachable, the perfect foil, as it turned out, both to Skilbeck's intimidating hauteur and to his sometimes over-ambitious exuberance. He was, moreover, the student's inexhaustible guide, philosopher and friend, one who dedicated many an unseen hour to sorting out individual and Union affairs.

The *Journal of the Agricola Club* (Wye College Magazine) in December 1947 contains Principal Dunstan Skilbeck's first 'College News', much in the same vein as one might expect from a headmaster on Speech Day. In it, he spells out the aims of the Governing Body, to expand student numbers to 200, including 80 women. He makes encouraging noises about the resurgence of healthy student activities, highlighting that 'a good deal of tolerable rugger, adequate cricket and rather questionable tennis' had been played. And even a College pack of beagles had been formed.

The main event of the year had undoubtedly been, in Skilbeck's view, the Commemoration ceremony in June to mark the 500th anniversary of the founding of the College. This provided the opportunity to invite many influential people to participate in the proceedings, such as the Archbishop of Canterbury,

Principal Dunstan Skilbeck on his round of the farms.

the Chancellor of London University and the then Minister of Agriculture. Also included in the ceremony was the presentation to the College of the original fifteenth century statutes, which had lain for centuries in the library at Merton College, Oxford. Such activities were 'meat and drink' to Skilbeck. They attracted the influential figures of the day, they guaranteed widespread publicity and they helped to create the impression of a fine academic institution, acutely aware of its tradition and intent on ensuring that cultural as well as academic standards were to be of the very highest order.

Mention in these College Notes also records the welcome news that the University of London had created two new Chairs at Wye, one in Horticulture and one in Agriculture. The former was to be filled by Professor Miles from the Advisory Service and the latter by Professor Cooper, about whom Skilbeck commented:

> Born of Scotch (*sic*) New Zealand farming stock, he comes to Wye with a knowledge of antipodean and of British agriculture, particularly in matters of animal husbandry, which must be hard to rival. Under his expert guidance the College dairy herd and the 1ˢᵗ XV should both show an outstanding increase in performance (for those who are not already informed on the subject, M.M. Cooper was captain of rugger at Oxford in 1936-37).

Thus the scene was set for Mac Cooper to take up his first professorial appointment for the academic year commencing October 1947.

7

WYE

The Coopers arrived in England in July 1947, after a five-week voyage across the Pacific and Atlantic oceans via the Panama Canal. Mac was a seasoned traveller and probably impatient after the first few days to get there and get on with his new job. For Hilary, of course, it was the trip she had been dreaming about for years and for the young girls, Barbara aged nine, and the twins a couple of years younger, it was the greatest adventure that had ever happened.

Hilary's parents had prospered as the export trade in motor cars and her father's related insurance business expanded. They had re-located back to Boars Hill, on the outskirts of Oxford, where they had purchased Wootton Close, a large house with spacious gardens. Florence Mathews, Hilary's mother, by now employed domestic staff, thus enabling her the time both to do local good works and to entertain generously. She was a formidable woman with a powerful personality and with considerable artistic talent in painting and pottery. She also moved easily within the social class of this exclusive neighbourhood and whilst keenly looking forward to the arrival of her beloved daughter and grandchildren, she also probably felt some apprehension as to what they would be like. Furthermore, in her early fifties, she did not take kindly to the idea of being referred to as 'Granny' or 'Grandma', so the alternative of 'Mops' had been agreed, whilst Wilfrid, her genial and somewhat long suffering husband, took the male equivalent of 'Pops'.

For the children, their new environment was a remarkable change from their wooden bungalow with a tin roof in Palmerston North. It never seemed to get dark in England (double summertime) and they loved to rush, barefooted, about the house and gardens. But it was difficult for them to adjust. As well as their grandparents, there were a host of other relatives, and even a great grandmother who was approaching her ninetieth birthday. Everyone spoke in a different way and their kiwi accents were laughed at by all and sundry. They had to start getting used to wearing shoes and learn not to call their swimming costumes, 'togs'.

After the initial excitement had died down somewhat, Mops was anxious to try and introduce a little discipline. It had come as a shock to find that there was so much to change in the manners of these little girls from the other side of the world. But by the time they moved to Wye a few months later, it is doubtful whether she had made much headway.

In a letter from Marygold Rix-Miller's mother to her daughter in New Zealand dated 10 August, she says:

At last we have seen Hilary, the girls and Mac. I haven't had a private talk with Hilary; she came with the three New Zealand whirlwinds and Mac. She said how well you looked and that James was a winner and how Mac admired Hugh. A pity for your sake that they have left. The three little girls told me that they don't like it over here and they want to go home . . .

The family moved to Wye in September, initially to a house in the village, but then to Amage farmhouse on the outskirts. This had been the house on a 90-acre farm, belonging to a trust, with the land being sub-let to the College,

Hilary's parents, Florence and Wilfrid Mathews (Mops and Pops).

who by now farmed over 600 acres. The house was a delight, dating back originally to the fifteenth century. There was plenty of space and a large garden within which the children could play. But adjusting to their new school was a bit of a culture shock, although fortunately the excellent primary school in Palmerston North had ensured that they were ahead of their age group, which made it easier to adapt and integrate. A pony called Mermaid and a golden retriever christened Kiwi were, however, soon added to the family, and even the 'three whirlwinds' began to think that life in England might not be too bad after all.

For transport, so essential in rural Kent, Hilary's father had managed to procure for them their first car – a black Morris 10. He had kept Hilary's driving licence safe for her but Mac had to pass his test. He proceeded to drive himself around Wye displaying his 'L plates' in preparation for the forthcoming examination, usually unaccompanied. Whilst he passed, driving seemed not to come easily to Mac. He neither understood the mechanics of the machine nor the speed at which one should move from one gear to another. Besides he was far more interested in looking over the hedge at the passing farms than he was in watching the road. At that time, the girls were always secretly relieved when their mother took the wheel. When they eventually grew up, they and other passengers were always quick off the mark to volunteer to drive so that Mac could watch the countryside in peace.

The new Professor of Horticulture had been installed for a few months before Mac Cooper arrived and the Agrics. were apparently concerned that their man had not yet materialised. There was much speculation as to whom it might be. It fell to Peter Wilson to be the bearer of the glad tidings that they were to have a New Zealand rugby star to be their Prof. Skilbeck had told him to spread the news before he posted it on the notice-board.

Peter describes Mac as a gifted speaker.

He always spoke without any notes, although his lectures were well prepared. He was a tall, lanky individual, with a mop of grey-brown hair, which he used to push back over his face from time to time. Unlike most lecturers, he never spoke from the rostrum or behind the lecturer's bench, but instead preferred to sit in a somewhat ungainly manner on the student's side of the bench, which gave him close physical contact with his class. His research-based lectures always had a practical edge to them. One always felt he was on the farmer's side, not merely delivering a scientific oration without relevance to the realities of agriculture.

As might be expected Mac approached his new job with tremendous energy. He was accustomed to students in New Zealand and the ones he found at Wye had the same sort of enthusiasm, many of them being like himself, ex-servicemen. Apart from planning the content of the teaching programme and lecturing himself, he was particularly keen to influence the farm operations and initiate some appropriate research work.

Another aim was to become involved in the sporting activities, particularly rugby, where, with his reputation, the students dwelt on his every word. John Fraser, an undergraduate in the early fifties and later to become a Governor of the College, remembers his first sight of 'Prof. Cooper', surrounded by admirers, kneeling in the quadrangle, illustrating a three-quarter movement by drawing with chalk on the flagstones.

Mac also wanted to get to know the local farmers, not only to gain their support but to persuade them that his students had much to learn from seeing their systems operate and discussing ways in which they might be improved.

Mac's relations with Skilbeck were cordial enough at first. However there were already signs of tension between two men with strong views and very different senses of priority. One of the most obvious, and probably the one which caused most discussion in the village pubs, was the Principal's continued insistence on riding around the estate, issuing orders to the farm staff. This caused confusion, as well as excuses for half-finished tasks. It might have amused the locals, but it irritated Mac to distraction. Whilst certainly a supporter of the College traditions, Mac also found it rather difficult to share Skilbeck's enthusiasm in upholding the stuffier aspects of some of the more esoteric activities, such as formal processions, the wearing of gowns, graces in Latin and 'High Table' for staff in the dining hall. Mac was altogether more down to earth, in keeping with most of the students, and therefore popular because of it. Not that he was in any way disloyal to Skilbeck, it was simply not his way and more often than not, rather than take issue in the Senior Common Room, he would bury himself in *The Times* crossword.

He was not beyond orchestrating practical jokes, however, no doubt with the aim of pricking a few inflated egos. There is little doubt that Mac had a hand in what took place at the Wye Fellowship ceremony in 1949. It had all the hallmarks of a Massey 'send-up'. The award of Fellowship was presented to individuals who, either whilst at the College or later in their careers, achieved eminence in their particular field. On this occasion there were three such people, one of whom was Sir Edward Hardy, a former Chairman of the Governing Body. As was the custom, a senior member of the academic staff was nominated to 'sponsor' and introduce each prospective Fellow to the assembled audience of local dignitaries. Normally the ceremony was reported as a rather solemn and formal affair. On this occasion, Sir Edward's sponsor was to be Mac Cooper. The tone of his introductory speech was slightly unconventional, including information he had allegedly gathered from the oldest inhabitant of Wye that: 'as a naval officer at Trafalgar during the Napoleonic wars, Sir Edward had been involved in some kissing incident. Then, as an ex-serviceman he had come to Wye to study religion and indulge in amateur theatricals.' The best was yet to come. The President of the Student's Union, Tony Orchard, was then rudely interrupted during his speech of congratulation by the arrival on the platform of three African Chiefs in full national costume. They duly proceeded to present the hapless Fellows with various awards, such as melons and ceremonial swords. Whether the Principal saw these proceedings quite in the same light as the (by then) delighted audience, is not recorded.

Apart from these light-hearted diversions, there also started to appear some more serious differences of opinion between the 'Prof.' and the Principal. This is best illustrated by reference to an article in the *Kentish Express* reporting a talk given by Skilbeck to the Ashford Chamber of Commerce towards the end of 1947. In it, he praised British farmers for their wartime efforts to boost food production. He went on to say that:

> . . . farmers were tired, they wanted to get back to producing livestock, our natural product, but they were being urged by Government to further increase output by 20 per cent, so that the country could save dollars by importing less wheat and beef. But although the Government promised financial support, where was the manpower to come from when German prisoners were being repatriated? The costs of this increase in production could well outweigh the benefits, especially if world food prices were to fall. What agriculture needed was a period of consolidation and stability; farmers needed to be properly on their feet before they attempted to run.

This opinion came as a bit of a shock to Mac Cooper. Food was rationed and he saw it as the Government's responsibility to encourage increased production as quickly as possible. Further, he knew that this would not be achieved if farmers were to return to the pre-war policies of extensive livestock farming.

Within the context of Wye and the College farms, Mac's prime interest was in the development of the dairy herd. With the establishment of the Milk Marketing Board and the guarantee of a regular monthly milk cheque, dairy farming was expanding. Fresh milk was also seen by Government to be a priority product for the population, especially children, who through free school milk schemes were encouraged to drink ever-increasing quantities. Here was an enterprise which enabled Mac to focus on improved methods of production as well as demonstrating the possibilities which well managed grass might have to offer. Even in 1948, he was advocating the need to expand grass growth by increased top dressing with nitrogen, extending the season by controlled grazing, and improving the quality of winter forage through more efficient methods of conservation. All of this was with the objective of reducing the reliance on purchased concentrates geared to lowering the costs of milk production. His farmer audiences received this evangelism with some suspicion. They preferred to see the results develop on the College farm before they deviated very far from their traditional approach. After all, concentrates might be more expensive, but they were also more reliable than conserved forage. What might be possible in New Zealand, with a ten month growing season, might not work quite as well during the six months cold and wet of a Kentish winter.

The College dairy herd became one of the first major development projects under the new regime. The proposals, however, were not without some criticism from the local conservationists. Christened 'Cooper's Cow Palace', an impressive set of new buildings was erected, consisting of four covered yards, each designed to house fifteen de-horned cows, a two level milking parlour, calving pens, a dairy and a field laboratory all under one roof. Dutch barns, two silage pits and a weighbridge completed the unit. Each yard contained a central feeding passage fitted with individual yokes, so that feed could be rationed and recorded for experimental work. This system also allowed concentrates to be dispensed outside the parlour, enabling a quick milking routine to be followed.

The Ministry of Agriculture provided some of the capital cost, although this did bring with it an unwieldy set of planning regulations, delay and red tape. One example was that the roof trusses ought to be obtained 'second hand', even though they would cost more than if purchased new. The Ministry architect also wanted to crown the building with a four-faced clock, presumably as a symbol of modern precision milking. Such diversions to getting on with the job are likely to have led to some added frustration for the person after whom the building was nicknamed.

As the project took shape, one or two of the more traditional neighbours were alarmed to see some of the permanent pasture being ploughed up and re-seeded to increase its productive capacity. Even a few ancient hedges were ripped out to increase field size and manoeuvrability for machines. One field of eighteen acres ended up with four rather than seventeen corners! But whatever the criticisms, it was generally acknowledged that, on completion, the unit was one of the most modern and best equipped in the country.

The aim of this very substantial investment was to demonstrate the theories which Mac Cooper had been expounding. Unless farmers could see for themselves the potential that grass and conserved feed had to play in lowering costs without reducing milk yield, he would not get them to change their outlook. At the same time, of course, this was to be a teaching site for students and also the resource within which more scientific feeding and management research could be undertaken by increasing numbers of post graduate students.

An article in the College journal explained that the sixty-cow unit was to be operated by three men, covering their own relief and holidays. The author pointed out: 'when one remembers that a skilled dairyman, after allowing for holidays, days off and overtime, costs not less than £350 per year, then the importance of saving on this item is fully realised.' He went on to say:

We cannot expect, with a surplus of milk being a not too distant prospect, to average 2*s*.10*d*. per gallon [equivalent of 3.1 pence per litre] for very long. Perhaps the time when milk will be 2*s*. a gallon [2.2 pence per litre] or less is not very far away and the College is anticipating this in its efforts to reduce those two major items of cost in milk production, labour and feeding stuffs.

Although the unit was to have a bullpen, it was not intended that it should be used in the foreseeable future. The College would be co-operating with the Milk Marketing Board in its new artificial insemination scheme, using fresh semen from a stud of promising Ayrshire bulls standing at the local centre at Whiligh.

AI had just begun to operate commercially on a localised basis, following extensive trial work by Hammond and his co-workers at Cambridge. This scientific breakthrough caused much discussion both inside and outside the farming fraternity, before operating licences (strictly under the Ministry of Agriculture control) were issued. Most of the original opposition came from the more extreme theologians, declaring that such practice was 'unnatural' and against the scriptures. Apparently a number of bishops were much exercised about the topic during a debate in the House of Lords. Before AI became widespread, farmers relied on natural service to get their milking cows back in calf. Most farmers either hired or purchased a bull from a pedigree breeder, based on a variety of dubious claims by the vendor. Not surprisingly, genetic progress in terms of increased milk production was very much a 'hit and miss' affair. Cooper now had the resources to make things happen on the dairy farm and he fully intended to use this as a launching pad to get his messages across to anyone who was prepared to listen.

Not all his senior colleagues, however, shared his enthusiasm. Understandably, the more scientific disciplines were intent on building up their reputation by the more conventional method of carefully designed experimentation followed up by publication of the results in the appropriate journals. This was especially true in the expanding Chemistry department as Louis Wain began to attract attention as a leading light in the research field. Whilst Mac's approach might be described as research, it was very much more applied than fundamental. Some argued it was more of a practical demonstration rather than a series of properly designed experiments. Mac was certainly more at home with this sort of work. He was happy to write articles and talk about the issues but he didn't relish the detail and the discipline of producing scientific papers. Some of his colleagues, therefore, questioned whether what they regarded as development work really helped the College's quest in seeking to advance its scientific credibility. In addition, there was the inevitable competition for limited financial resources. The 'Cooper Cow Palace' had cost a great deal of money and there were rumours that other investments on the farm side were in the pipeline. Other departments were anxious to improve their facilities and the funds were limited.

Finally, there were some untidy lines of responsibility as to where the buck for the farming activities actually stopped. A 'Farms Committee' within which Mac Cooper was the leading light reported to the Governing Body, which in turn took the major decisions. Mac, however, was not a member of the Governing Body. Moreover, James Wyllie, a rather dour Scot with a reputation for meticulous records but few exciting ideas, produced quite critical reports on the farm enterprises, which went direct to the Governing Body, without (much to Cooper's annoyance) even passing over Mac's desk. Wyllie had been head of the Economics Department and although now taking a back seat, he was still held in high regard by many of the senior staff and Governors.

Whilst these matters were irritating, they did not, at this stage, cause Mac to lose much sleep. It seemed to him that he had the backing of the Governors, especially the farmers amongst them, and in any case he had never taken much notice of academic in-fighting. It also happened that soon after their arrival at Wye, Mac and Hilary began to meet many of the surrounding farmers and their wives. Some of these friendships lasted all their lives. As well as much social activity and generous hospitality, the men enjoyed many hours of spirited debate and argument. More importantly for Mac, it allowed him to escape from the internal politics of the College and gave him the confidence to stand up for what he believed to be what local farmers wanted. This group of friends presented Mac with a silver ashtray when he eventually left Wye, inscribed: 'This once we have all agreed! Jack and Dick Merricks, John and Bob Montgomery, Dick and Don Cooke and Reg Older.'

In order to 'make things happen', both in the student curriculum and on the farm research side, Mac was constantly on the lookout to recruit staff who shared his enthusiasm for grassland farming. One such individual was Canada Davis; not his real Christian name, although he had been born in that country. When introducing himself to the Director of the Welsh Plant Breeding Station, he explained that his name was Davis. The reply was that as the Station already had three others with that name, he would henceforth be known as Canada. It stuck ever since. Mac had met Canada in 1948 during a student tour which included the Grassland Research Station at Stratford. A quiet and precise man, with his feet firmly on the ground, Canada had previously been a part-time lecturer at the Royal Agricultural College at Cirencester. Most of his students had also been ex-servicemen, and one had even held the rank of general! Over the years he effectively became Mac's right hand man, involving himself with experimental work, teaching, organising student tours and generally ensuring that loose ends, of which there were many, got tied up. Usually with baler twine!

Canada recalls the regular farm meetings held in Mac's tiny, smoke-filled office every Saturday morning. There was no agenda, no minutes, just a monologue describing to everyone what was going on and what the priorities were to be for the following week. Also the responsibilities for those staff delegated to talk to visitors about the research projects which were underway. Most weeks, groups of local farmers or overseas academics would appear, sometimes without notice, and expect to be shown around. It was all rather haphazard, but good-natured and dynamic. Things were always happening. Life was exciting.

Canada also remembers the 'Pig Club'. Again this initiative is recognisable from previous years at Massey and Cambridge, when Mac helped McMeekan dissect pigs. Membership of the club at Wye was open to able-bodied staff who were willing to lend a hand in feeding the pigs, usually early in the morning or late at night. In return for their labours, joints of pork were duly shared out after the animal had completed its experimental programme, been slaughtered, and dissected. Mac, of course, took his turn and on one occasion was tipped half a crown by an enthusiastic overseas visitor who mistook him for the pigman. Needless to say he pocketed the money gratefully with a due doff of his cap.

Meat was a luxury and the club attracted some other notable members, including Louis Wain, the head of the Chemistry Department. Mac trusted no one other than himself to dole out the meat, so ensuring everyone had their turn of the best and the worst cuts.

Pigs had been a favourite species from Massey days. They held much greater potential to explore the possibilities of faster genetic improvement than did dairy cows. Work at Wye, therefore, soon included a bona fide pig testing station, set up to test eight boars per year, based on Danish standards. As new buildings materialised, a research programme to determine genetic variation in pedigree Large White pigs, coupled with evaluation of existing techniques in progeny testing, commenced, as did various feeding trials to try and improve feed conversion efficiency.

Mac Cooper lecturing at Wye. (John Topham, Ltd)

Poultry were also an enterprise which provided postgraduate students, interested in genetic improvement, the scope to measure progress quickly, because the breeding cycle was rapid and improvements could easily be measured through egg production. One little anecdote concerned a breeding trial with Rhode Island Red cockerels. Mac had invited the local pedigree breeders' club to come and assess these birds in respect of their daughters' laying ability. Much to his satisfaction, the breeders managed to rank them in exactly the opposite order compared to the actual results. Whilst perhaps a frivolous example, this was an area of increasing concern to the breed societies, in that the science of genetics was beginning to challenge their traditional market place. Until now, pedigree breeders had enjoyed a virtual monopoly in selling males for breeding to the commercial producer. With the arrival of AI and progeny testing, those with a little vision could see the writing on the wall. Mac could also see that this was to be a subject which would run and run. He also began to speak authoritatively on beef production and the scope that there was for improved breeding and selection of stock based on recorded performance, rather than external appearance. This didn't go down too well with the pedigree beef breeders either. He encouraged the controversy. He was able to get farmers to think and to argue amongst themselves. His confidence was increasing and he was enjoying the experience.

It was not until September 1949, however, that the big breakthrough came. Mac was thirty-nine and had been at Wye almost two years. He had made an impressive, if at times controversial start, although his impact had been confined largely to technical matters and references to what might be possible if farmers were to adopt a more positive attitude to grassland management. He hadn't as yet made the national headlines, but this was about to change.

Alan Stewart was the second Rhodes Scholar to come to Oxford from Massey shortly after the end of the war. He had been one of Mac's students as well as the College rugby captain, so he had a great deal in common with the Cooper philosophy. Alan recalls that after finishing at Oxford, he was having difficulty completing his thesis. Not only were his tutors unhelpful, but he was also recovering from a broken leg, the result of a rugby injury. The Coopers had invited him to stay at Amage, so that he could recuperate and at the same time take advantage of the help which Mac could provide to finish his thesis. He stayed for several weeks. One weekend he remembers that he and Hilary were involved in the incident which led to Mac Cooper becoming a national figure.

He wrote a paper (somewhat unusual for him, he preferred off-the-cuff lectures, at which he was adept) in which he criticised British agriculture and said that the more he saw of it, the more he thought that it was only at half cock. The agricultural press could not believe its luck and thereafter he was known as 'Half-Cock Cooper'. When he went to Newcastle this was improved upon by a journalist in a poultry magazine, I think, who converted it into 'Half-Cock of the North Cooper'. Where Hilary and I came in was that Mac read his paper to us in the front room in the house. Hilary subsequently went to sleep and as an aggressive colonial I couldn't see anything wrong with it.

Alan Stewart was, himself, no stranger to controversy. After a distinguished career in the Royal Navy during the war, he returned to lecture at Massey. Following a visit to New Zealand by some senior members of the Milk Marketing Board, he was invited to return to England to set up a Consulting Officer Service. His job was to persuade pedigree breeders that progeny testing was a more reliable way to evaluate the potential of a young bull, than a reliance on family history. By the introduction of a supplementary type classification system, administered by the Consulting Officers, Alan was able to make considerable progress with the 'farmer panels' who were asked to inspect the test bulls' daughters on the farm. It seemed the MMB couldn't find anyone in England to undertake this somewhat sensitive task. After returning to teach at Massey, he eventually became Vice-Chancellor of the University and received a knighthood in recognition of his services to New Zealand agriculture.

Mac's paper, to which Alan referred, was presented at the meeting of the British Association of the Advancement of Science in Newcastle in early September 1949. The national press, not just the agricultural journals, seized on the story, much as they would do today when a prophet from foreign parts has the temerity to tell farmers, politicians and academics that they are living in cloud cuckoo land. Newspaper headlines like 'Our Expansion Could be Doubled', 'Heretic Scientist', 'Scientist Condemns Huge Farm Subsidies', spelt out the message that British farmers were feather bedded and that subsidies merely kept prices too high and helped to keep inefficient farmers in business, when really their places should be taken by men who could move with the times.

He was more specific when he talked about dairying.

The level of British milk yields was too low to be tolerated, due to a variety of controllable factors, such as deficiencies in feeding and management; poor inherent quality of stock through indiscriminate or unsound breeding; and high wastage as a result of disease.

A national average of 700 gallons (3,180 litres) and more is not an unreasonable goal in a country with educated farmers. It would mean that fewer cows and less land could satisfy present needs. Labour and feeding stuffs would be freed for the production of mutton and beef.

The extent of the media interest in his pronouncements was a great surprise to Mac, as well as to his colleagues back at Wye. It certainly provided publicity for the College and delighted the students, who felt a sort of reflected glory in their Prof.'s outspokenness. However, it further alarmed some of the traditionalists who feared a farming backlash from the establishment. Some felt Mac was becoming a loose cannon and wondered what might be the next target on his list. They were not convinced that he was furthering Wye's reputation in the eyes of those who mattered. It may well have been at this stage that Skilbeck, perhaps under increasing pressure from some of the Governors or other senior staff colleagues, tried to persuade Mac to tone down his approach. It has to be said that there is no evidence for this, but it is clear from private correspondence with his relatives back in New Zealand, that relationships between the two men started to deteriorate significantly in 1950.

Such a potential confrontation would have come as no surprise to Mike Soper, the author of *Years of Change*, Oxford academic and the doyen of the Oxford Farming Conference. In some private correspondence, he relates his experience, when he went down to Wye to discuss a possible job with Skilbeck when the College re-opened.

Both Mac and I were on Skilbeck's short list of 'possibles' for two jobs at Wye in 1946, one of which was the new chair. When I went down to talk to Dunstan about it, he didn't make clear which one he had in mind for me. In the middle of the interview he threw a vast tantrum with some unfortunate RAF subordinate on the telephone, and in an instant I decided that he would be very difficult to work with, so I turned him down, and decided to go to Oxford instead, as I had an offer from Geoffrey Blackman on the table.

If it hadn't been for his volcanic behaviour, I might well have accepted the offer at Wye. I've often wondered how Dunstan and Mac got on. I'm sure there must have been some fairly monumental rows from time to time, as they were both highly strung characters.

In a letter home to his sister Madge, in September 1950, Mac seemed to be at low ebb, with frustration constantly bubbling below the surface.

It is just a year ago that I achieved such notoriety from my British Association talk when I accused British farming of being at half cock. The more I see of it the more do I re-affirm my point of view. The farmers are completely spoiled by the prices they get and the great majority, in consequence, have no incentive to be progressive.

In the same letter he talks of showing senior researchers from New Zealand round the College farm and records that they seemed very impressed. However, it is clear that all is far from well.

Personally I am rather disappointed with progress so far. There is a tremendous amount of passive resistance to change and I, on the other hand, am rather impatient. My greatest stumbling block is this man Skilbeck who is Principal. I have tried my best to be loyal to him, but unfortunately he is thoroughly jealous of me.

If I weren't so interested in what I was doing, I could not have lasted twelve months. As it is, I shall take the first reasonable opportunity I have of getting away from him. I cannot stay at Wye and maintain my self-respect. If only I had the capital I would go farming on my own account. At least I would be answerable only to myself and would be able to cash in on the inflated prices being paid to farmers.

Whilst it must be remembered that this is in the context of a private letter home, nevertheless it is a rather surprising outburst, at a time well before Mac actually left Wye. It seems that one of his difficulties was that he found it very hard to share his problems with anyone else, including his wife, so it all tumbled out in letters home. Maybe he felt he ought to be able to sort these things out himself; they were professional matters, only he could resolve them.

The staff would have preferred that this rift between the two personalities be confined to the Senior Common Room. However, it soon became obvious that separate camps had begun to develop within the College. Jerry Rider, a student from a non-farming background and now one of Britain's most successful dairy farmers, remembers the atmosphere when he went to Wye in the early fifties. 'You soon knew your place: the right school, debater and beagler – a Prin. Man. Any school, beer and rugby – a Prof. Man.'

Life of course went on, whatever the internal politics. Mac continued to be in constant demand to talk to academic as well as farmers' meetings and to appear as a speaker at the ever-increasing number of agricultural conferences. He rarely turned down an invitation. His style belied the nervous energy he put into these platform performances. The audience never ceased to be amazed by the fact that he never referred to a note, but seemed able to talk off-the-cuff for forty minutes, often including detailed facts and figures. But behind the scenes, he had prepared very carefully and for an hour or so before the event, he was usually 'incommunicado'. This took a lot of effort and when these meetings came in quick succession, sometimes three a week in different parts of the country, and on top of all his other commitments, he would become completely exhausted, sometimes retiring to his bed for two or three days to recover.

His main topic continued to be the grassland theme of increasing productivity, but he also began to launch more attacks on pedigree breeders. In a provocative address to the National Veterinary Medical Association in October 1951, he deprecated the 'illogical cult of ancestor worship and the adherence to the cow family system of nomenclature'. He opined:

At least genetics has progressed to the point where it can explain that, when an animal is used for breeding, it transmits only a sample half of its inheritance and this may be good or bad, according to chance. Yet breeders will complacently point to this outstanding animal, which is probably provided with some high sounding name, that is somehow preserved into the generation under consideration, in the belief that they are safe-guarding their breeding plans.

The cow family cult was of the same order. Animals had been retained for breeding in preference to more productive cows, simply because they happened to have fashionable family names.

This was just as logical as attributing all the good points of one's own daughter to a great-great-grandmother-in-law.

Mac did concede that in the meat sector there was virtually no method of recording performance other than by eye-appraisal. Certainly this was true with beef and sheep, prior to the introduction of supervised weight recording programmes. Pig recording was beginning to get off the ground and the poultry industry was fortunate in that egg production could easily be measured.

Milk recording, however, now offered a golden opportunity to gather information from which to make informed, rather than ill-informed, decisions as to which cows to breed from. This coincided with the work on progeny testing started by the MMB and dairy bull evaluation, based on contemporary comparison of daughter records in different herds. This was to revolutionise cattle breeding in the next few years. In an audience largely comprising vets, these broadsides against the traditional breeders went largely unchallenged. However the press had a field day and Cooper came under another storm of criticism which kept the publicity machine chugging along nicely. In December of the same year, he had a go at the recent Government Annual Price Review, declaring:

It was just as fickle as a young woman's heart. There was a concept that taps could be turned on in Whitehall to decrease milk production and increase beef.

But farming needs to be planned at least ten years ahead. There was too much security for the 'B' farmers, and those who just pottered about without really farming, and too little for those who really did the job. Neither could he understand, that after 12 years of food rationing, why rather more than 40 per cent of agricultural land, other than rough grazings, was still in permanent grass.

It was as if Mac Cooper was becoming the thorn in the side of the establishment. His was the opinion the media sought whenever there was a controversial issue to debate. He was becoming increasingly well known

Mac on the farm and on the platform. Wye, circa 1949.

throughout the country and also well respected for his no-nonsense opinions, whoever they might offend.

Personal relations between the 'Prof.' and the 'Principal' at Wye had remained very strained for most of the previous year and although Mac, in usual style, battled on regardless, Hilary had suffered a nervous breakdown. Whilst there can be little doubt that the internal politics at Wye, and Mac's withdrawing in on himself, were the main causes, an additional factor would have been the long and worrying illness which kept Barbara, their eldest daughter, away from school for many weeks. In a letter home to Thelma and husband Joe at the end of the year, Mac described Hilary's condition some nine months previously, when the unpleasantness between the various staff groups had been at its height. After a period of convalescence in Oxford, she had returned to Wye but suffered a relapse in the Easter vacation. He wrote:

> She appeared to be as bad as she was before and getting rather desperate about it all. Unless you have had some experience with a nervous breakdown, you just cannot appreciate what it all means. Physically there is nothing to be picked up and yet there is real illness, which cannot be conquered, as some people think, by repeating a cliché about mind over matter.

He went on to say that she gradually improved as the spring approached, being very responsive to weather conditions.

> She feels the cold dreadfully. With bright sunshine she is 100 per cent, but with miserable weather, I have got to provide the sunshine. It has been a bigger strain than I have ever shouldered before, but it has been a very worthwhile effort for so much is at stake. To some extent, Hilary's illness has accounted for my silence [he was an intermittent correspondent]. I just haven't had much heart for writing, at the same time I must admit to being a bit 'cooperish' about it.

One might reasonably define 'cooperish' as 'a withdrawal into one's shell under pressure and to keep one's thoughts to oneself.' It would seem to be a highly heritable trait! He concluded in his letter to Thelma that his relations with Skilbeck were now much better than they had been for a while.

> I had a showdown with him during the year and had my position reviewed by the Board of Governors. I know just where I stand and have had my authority properly established. I couldn't have carried on as things were. I blame this situation for Hilary's trouble more than any other factor.

Life at Wye in the mid fifties continued to gather momentum. Thanks in large measure to Skilbeck's ingenuity in attracting funding from all manner of University, Government and commercial sources, new laboratories, lecture theatres, and improvements to the existing fabric of the College were constantly in the planning or building stage. When the opportunity arose, acquisitions were made of additional land for the farms as well as accommodation in the town suitable to house an increasing number of overseas and post-graduate students. Inevitably, such a policy of expansion did not please all the local inhabitants.

Wye had also built up something of a reputation overseas, with regard to its involvement in agricultural education since the turn of the century. Many early graduates had either joined the Colonial Service or worked abroad managing plantations and estates. As the old Empire administrations were replaced, Government policy was directed towards helping these emerging and newly independent countries to develop their internal education systems based on British experience. A number of senior staff at Wye undertook visits to the Sudan and to such countries as Nigeria, Uganda and Rhodesia to evaluate the requirements of the establishments which were being set up. This was linked to a London University Special Relationship Scheme, whereby the curricula were approved and examiners appointed. The Inter University Council frequently consulted Mac Cooper and he often helped interview applicants for overseas posts. As a consequence, he made many trips out to Africa, advising on the content of courses, acting as an external examiner and helping to identify the most appropriate research activities for that particular country. His, and other colleagues' commitment to these worthwhile contacts brought the reward of widening the College's reputation. Over the succeeding years, Wye became popular as a destination for overseas students and post graduates, as well as an obvious centre for international meetings involving senior agricultural directors from throughout the Commonwealth.

The fifties also heralded a significant change of emphasis within the Economics Department. Under James Wyllie the Department, whilst highly regarded locally, had for some years concentrated on detailed cost recording and survey work. It had become tied to convention, rather than seeking out new methods of analytical evaluation. One of its main functions was to work with the Ministry in farm income and cost investigations as part of the national programme used in the Annual Price Review negotiations. So there was a quasi-Governmental atmosphere which might have been a limiting factor in stimulating new ideas. With the appointment of Gerald Wibberly in 1954, the whole emphasis changed. Farm management became a subject in its own right and some fundamental questions began to be asked about the content of the academic courses as well as the activities on the farms.

One of the new men of this era was Ian Reid, who as a Farm Management

Liaison Officer arrived at Wye in 1953. He was to be the linkman with the National Agricultural Advisory Service (NAAS), whose staff provided free advice to farmers. One of Ian's jobs was to run courses for farmers, land agents, bankers, etc., designed to evaluate the financial consequences of adopting the largely technical advice which was being dished out to all and sundry. As he became more involved with the students and started teaching, there developed areas of quite strong disagreement with Mac Cooper who, in Ian's view, was very much a 'production man' who didn't really understand economics.

Ian arranged visits for the students to a number of farms which appeared on the surface to be run down and showing all the signs of poor technical performance. However on closer examination, they recorded reasonable income with virtually no costs, thus returning a very respectable profit. On another occasion he took them to a beef enterprise where big steers were sold at three years old. Without knowing that they went to the kosher beef market and commanded three times the market price, such a system would appear as hopelessly uneconomic. Ian's main point was that the College farm was designed to be, on Mac's own admission, a microcosm of British farming, showing examples of enterprises at high performance levels. This might be acceptable for demonstration purposes, but it was misleading as an indicator as to what might be the best choice on a practical farm under commercial conditions.

One of the problems which was coming to light with the introduction of costing, was the very low contribution some enterprise activities made to overhead costs and profit. Many farms had a few pigs, beef, sheep or poultry to help spread risks, but when their financial contribution was measured, they not only failed to cover the variable costs of production, but also involved additional labour, already in short supply. Getting rid of them not only increased profit and reduced workload; it released capital tied up in the stock. Ian Reid's thesis was that if prices and circumstances were right for high cost, high output farming, then farmers should go for it, provided always that they used their resources efficiently. But there were alternative strategies and one must be very careful not to draw conclusions on the basis of technical performance in isolation. There were plenty of examples where either the farmer had a specialist market which commanded a premium price, or where his cost structure was so low that he could make a profit at what appeared to be appallingly poor levels of output.

The renaissance of the Economics Department triggered a more far-reaching influence than just at farm level, important though that was. A survey showed that almost half the students felt that it was essential that future courses should include increased components of farm management, economics and marketing. Furthermore, it was feared that if this did not happen, future students might alter their choice of University. Already there had been an increase in the number of postgraduates applying to undertake management related research projects at the expense of the natural science specialities.

Student numbers were vitally important.

Thus followed another long battle between the proponents of the traditional science faculties, headed by Wain, against the new breed of economist led by Wibberly, better known as the Welsh Wizard owing to his undoubted debating skills, with Skilbeck sitting rather uncomfortably in the middle. By the time this confrontation got into full swing, however, Mac Cooper had moved on. It would have been interesting to see which side he would have supported.

After the 'showdown' with Skilbeck and for the rest of the time that Mac was at Wye, there grew a mutual understanding between the two men, although it was an uneasy truce. Within the College there remained the two camps, even to the extent that hosts of social gatherings and dinner parties were careful in issuing invitations, to avoid creating embarrassing clashes between the various protagonists. Whatever the rights and wrongs of this clash of personalities, there can be no doubt that each, in his own way, had made a major impact since the College had re-opened after the war. Skilbeck had the charisma to persuade those who held positions of authority and influence that Wye was going to be something special. Not only was he able to recruit able staff, he had a clear vision as to the sort of curriculum and establishment which would attract students from a relatively prosperous sector of the industry. He was aware of the changing market place and he would allow no one to stand in his way. Cooper on the other hand was a farmer's man, a pragmatist, less comfortable with the internal politics of the Senior Common Room. He was also over

Cooper and Skilbeck at the University ploughing match, 1953.

118

sensitive and resentful of the fact that he did not have the power to make decisions quickly. He tended not to talk about these problems to anyone and yet worried about them a great deal. He certainly felt that because of his own success, Skilbeck was jealous of him, whereas he might have expected someone with more humanity to be pleased that his hard earned reputation was linked directly with the College.

In fact the two men were reconciled when eventually Mac left to take up his appointment in Newcastle. In later years they came to realise that they had both been trying to build the same house using different bricks. But Mike Soper was right when he doubted whether the two of them would ever work successfully together.

By the beginning of 1954, Mac Cooper had been at Wye for over six years. He had known for some time that a move was unavoidable, if he was to continue to progress his career. There was only one way to ensure that he could avoid the risk of repeating the frustrations he had faced at Wye and that was to be his own boss. In the event three jobs came up at the same time. In a letter to Thelma, dated June 1954, he describes the sequence of events.

> I am sorry I was not able to give you an inkling of my change of job before the news got out to New Zealand. It all happened very quickly and I didn't bargain on it being cable news. Three months ago I did not have a thought that I may be moving so soon, but within a week I had the offer of three jobs – a new one in Ireland, which would have been a tremendous gamble; the Chair at Reading; and the Newcastle post. The Reading job was the easiest and superficially attractive but Newcastle gives me the greatest scope to develop my ideas and also the greatest freedom of action. It is about the most senior Chair in agriculture with the possible exception of Cambridge, which does not appeal to me in the slightest.
>
> Today Hilary has been getting out some snapshots of our New Zealand life. They have made me rather homesick for the old days and I wish I could recapture something of the simplicity of our lives then with the children at such a wonderful stage, like flower buds just opening up. There were some snaps of me in my younger days – a small boy with Trixie in front of the house at Mangateretere. That was one life, Wye has been another and now the prospect is something very different. I do hope we can find the sort of house we want so that we can make it a real home and give the children a background they can always cherish.

This letter home was also tinged with sadness as Mac had just heard of the death of his eldest brother Jim. He had, of all the family, been a 'tower of strength' for Hilary when Mac was away at the war. His most vivid memory of Jim was 'his walking down Te Awe Awe Street with a twin clutching either hand. He had such a wonderful capacity for understanding children and making them love him.'

There were other good reasons for the move to Newcastle. The main one was that the appointment was as permanent Dean of the Faculty, so there was no ambiguity about who was in charge. The Reading job had a rotating Dean, so that once the three-year term had expired, the same old problems of leadership would have surfaced again. Whilst the Newcastle job carried a higher salary, this was less significant than the perk of free University tuition for the girls. By now, Barbara had decided to study medicine and although she had set her heart on going up to London, Newcastle also had a School with an impressive reputation in this field. It was a little early for the twins to decide on their future careers, but free University tuition was a valuable option, should they wish to take it.

Leaving Amage and Wye, with all its stresses and strains, for the far north of England was going to be an upheaval for the family, as well as a journey into the unknown. For Diana and Cynthia, the thought of leaving all their friends behind as well as their pony (they had to live in Newcastle until their new house was built) seemed like the end of the world. For Barbara it would mean her remaining at school in Kent as a boarder, so that she could complete her final year and A levels. Hilary, on the other hand, was ambivalent about the move. She knew that Mac would never be happy until he was his own boss, but despite all the traumas of the past few years, it meant leaving lots of friends, a lovely home and also substantially increasing the distance from Oxford and her parents. But perhaps worst of all, for those unfamiliar with the North-east, Newcastle had an image of bleakness and coal dust, very different from the rolling countryside of Kent and the 'garden of England'.

8

NEWCASTLE : THE DOSE OF SALTS

Nowadays Newcastle is a University in its own right. In 1954, the School of Agriculture, King's College, although located in the city of Newcastle upon Tyne, was an integral part of the much older University of Durham, some fifteen miles to the south. Formerly Armstrong College, King's was growing rapidly as a multi-faculty campus, including a full range of pure and applied science courses, arts and also impressive medical and dental schools. The Agricultural Department had been founded in 1891 under its first Professor, William Somerville, and over the years had built up a formidable reputation in the North-east and elsewhere.

It was for Mac Cooper, however, a very different environment. Newcastle was a 'red-brick' establishment, with none of the traditions or the insularity of Wye. There were five thousand students pursuing all sorts of degree courses, so that the interests and activities taking place covered every aspect of University life, both at student and at staff level. The buildings were in the centre of the city, with a number of miles of built up area between the College and any agricultural enterprise. There were College farms, but the students had to be bussed to them. Finally and most importantly, agriculture was only one department (admittedly a noisy one) amongst many. The students were therefore less parochial and could more easily mix with contemporaries from other disciplines and backgrounds. Not that these factors were necessarily seen as either advantages or disadvantages, they were merely different.

Mac's predecessor, Professor R.W. Wheldon, had died unexpectedly in harness earlier in the year and the University Appointments Board, once they had made their selection, were anxious that their new man should take up his post with the minimum of delay. It seems that they had done their research, identified the best man for the job, and invited him to take it. No advertisements, no interviews – just get on with it.

Wye, on the other hand, seemed to be in no hurry to find a replacement Professor of Agriculture. No doubt following the difficulties recently

121

Armstrong College, circa 1904. Now Armstrong Building, University of Newcastle upon Tyne.

The original Agricultural Department, now occupied by the Architectural Faculty.

experienced the Principal would want to influence the Governing Body as to the future positioning of the farm responsibilities, before he commenced his search on their behalf. As usual in these matters, a compromise was agreed, resulting in Mac beginning his involvement at Newcastle in June 1954, which coincided with final examinations at Wye, as well as his commitment as an external examiner at Bangor, Aberystwyth and Reading. On top of that he had at least eight post graduate students at Wye who were in the process of completing their work.

To say that he was at full stretch would be an understatement. In a letter to Thelma, he says, 'Hilary and the children see so little of me these days and when I am home I appear to be in a dream.' The long summer vacation might have been expected to provide some respite, but this *inter regnum* period at Newcastle carried on until mid December, some eight months after he had accepted the job. In practice it meant weekly commuting between Wye and Newcastle, a round trip of over seven hundred miles. Fortunately he could use the train and the travelling time to catch up on administrative matters, but the unsatisfactory nature of the continuing responsibilities at Wye were a substantial extra burden which he could have well done without. Clearly his mind would have been preoccupied with the new challenges at Newcastle, but he would also not want to short-change his former students.

It was not until August 1955 that Bill Holmes took up his position as Mac's successor at Wye. Bill, a Scotsman, had been Head of the Department of Grass and Dairy Husbandry at the Hannah Research Institute, near Ayr, in Scotland. Unsurprisingly, he did not inherit responsibility for the farms, which rather irritated him. However it was obvious that the Principal was not going to allow any danger of a repeat of the problems previously encountered. Nevertheless, Bill's appointment, together with a number of other staff changes at around the same time, returned the College to a more unified front. The Skilbeck/Cooper clash had been debilitating and gone on for too long. According to Donald Sykes, who had joined the staff as one of the new breed of economists in 1952, 'People were tired of the in fighting. Life was to be a good deal smoother, if rather less exciting.'

The press welcomed the news of Mac's appointment at Newcastle. A Table Talk article by 'Pendennis', in the *Observer*, dated 23 May 1954, stated:

Well-deserved promotion comes this week to one of the most colourful and vigorous personalities in British farming, Professor M.M. Cooper.

At present he teaches at Wye College, part of London University, specialising in agriculture. In October he moves north from the gentle Kentish countryside to the sterner Pennine slopes to become Dean of Agriculture at King's College, Newcastle.

Northern farmers will find their candour well matched in Cooper, whose lanky figure, unruly hair and forthright manner make him the unstuffiest of dons.

Mac's own thoughts about the new job are included in a letter to Thelma, after he had moved up to Newcastle with the family to a terraced house within a couple of minutes walk of the College.

I am really thrilled with my job at Newcastle, which provides tremendous scope. There are two farms, Nafferton of 730 acres, which is mainly devoted to dairying, corn, potatoes and sheep, and Cockle Park which is 520 acres with cattle, sheep and some cropping. The latter is sadly run down and there has been very little in the way of new ideas there for many years. My intention is to develop it as a research unit. It has a tradition in research, for it was the first place in the world where basic slag was used as a fertiliser. There were some great names associated with it in the past – Somerville, Middleton, and Gilchrist, but latterly Cockle Park and the School of Agriculture, which is behind it, have suffered badly from inbreeding. My predecessor, Wheldon, had never been away from King's.

Nafferton will be a purely commercial farm and the aim there will be to streamline production. There is difficulty when one tries to combine research and teaching, of having to serve two masters and one eventually suffers. So I need to keep them as separate entities. I am fortunate that this is the only School of Agriculture in England with two farms. There are neither pigs nor poultry at either farm and one of my first jobs is to develop units of these. All told, I will have a £40,000 building programme, because the existing buildings are shocking and we must have more cottages for staff. The farms are not my only problem. The teaching is moribund and the student's day is packed with lectures. They take down notes to dish them back again in examination papers in the hopes of getting a degree. To think that people have considered that sort of thing is University education! The essence of it must be critical thought and an incubation of fresh ideas. Yes, life is going to be very busy over the next few years but it will be fun doing it and I shall be very much my own boss.

As you know three jobs came up at the same time. I chose King's because it was the biggest challenge. Reading would have been very pleasant in many ways, but it is too close to London and it is swamped by surrounding research centres. I have to create a research centre up in Northumberland. My third offer came from Southern Ireland. It was rather nebulous and it would have been fun for a few weeks, but I am certain it would have had me believing in fairies before very long. The Irish are delightful people entirely lacking in a sense of time and completely inconsequential. It is a grand place for a holiday but not for work.

Mac goes on to explain that they were hoping to move into their new house at Cockle Park, in the spring.

It is not large but very compact and convenient, which will be a blessing for Hilary, for though Amage was a very dear place, with a soul to it, nevertheless it was a housewife's purgatory in some respects. There is no garden and we will have to start virtually from scratch. I love gardening and I am going to enjoy planning it and provide a setting for our home. It won't be like gardening in Hastings though, because the winters are so very long. We won't be able to have grapes or peaches but small fruits such as raspberries and strawberries grow extremely well.

Most people think of Newcastle as a grimy industrial city. This is true of some parts of it, but where we are is really quite pleasant. The industrial part of Northumberland is mainly along the Tyne and once you get out of it, there is some of the most heavenly country. It reminds me more of New Zealand than of any other country. There is a rugged openness about it that does my heart good. It is, of course, completely redolent with history and right from Roman times till the 1745 rebellion was a 'no man's land'. You will see from the map how Northumberland angles up into Scotland and the border is not very far away from any part.

At Cockle Park there is an old fifteenth century pele tower, which is a relic of the border days. Its walls are massively thick and there are the marks of cannon balls on them. It was a rallying point when forays were made across the border by raiding parties.

The Pele Tower, Cockle Park. Formerly a residence for post graduate students.

He ends this long letter to his family in New Zealand by saying: 'I keep up my rugby interests and I am President of the rugby club. I haven't started coaching them, but will do next term, when I get more time.'

His arrival at Newcastle came as a shock for staff and students alike. His reputation went before him, although no one knew him personally. Clive Dalton, a local hill farmer's son, went to King's as an agricultural student in 1953, the year before Mac Cooper took over. After completing his PhD at Bangor, and doing some teaching, he subsequently emigrated to New Zealand to work in sheep research. A gifted communicator, and still possessing a rich Northumbrian accent, Clive is well known in New Zealand as a broadcaster and commentator, covering a wide cross section of interests. He recalls:

I wasn't really sure what was involved, as three years seemed a lifetime of study, so when 'aa gat haeme' all I said to my North Tyne farming friends was that 'aa was gannin te King's Colleege' (which many still called Armstrong College, where daft professors came from).

The first year was murder! It was back to school, but school in a hell of a hurry! Classes with big bright students from other departments, including medicine, to study the pure sciences. Lecturers had little time for whom they considered 'thick' agric. students who might be struggling a bit.

I was in my second year when Prof. Mac arrived from Wye College. We Northumbrians were a bit suspicious of anybody from 'doon sooth' which technically is south of Gateshead! (the town on the opposite side of the Tyne to Newcastle). But we had nothing to fear. He was a different creature to anything we had seen from those foreign parts who had queer accents, and who complained that they couldn't understand the way we 'taalked'.

It was an incredible experience from Prof. Wheldon who we never met, but knew by sight, in his bowler hat, spats and umbrella arriving each morning from the Central station. Prof. Wheldon's sudden death had little effect on us, as the man was a stranger to us lowly undergraduates. Then this tall, lanky, tanned New Zealander arrived at the lectern each week (initially he was teaching both at Wye and Newcastle) in his Harris tweed suit. We were a bit terrified at the start and you could hear a pin drop during his lectures. Harris tweed itself was a shock in a climate of dark blue or grey pin stripes.

Prof. had notes that he never used. He talked about far away places with strange sounding names like Ruakura, Waikato, Manawatu and Massey. He had a global perspective on things, and saw agriculture as the most noble occupation of mankind. He impressed on us that it was both science and an art.

The students and the press may well have approved of Mac's appointment, but it came as a severe blow to the incumbent senior staff, some of whom had been expecting to be approached themselves.

It was clear that the chief architect behind the scenes had been the Rector of King's College, Dr Charles Bosanquet, who had himself, gained considerable experience in agriculture. The Bosanquet family had owned Rock, a 2,000 acre estate near Alnwick in Northumberland, since 1804 and although under professional management, Charles remained involved in its operation, albeit at a distance. His own career had commenced in financial journalism after graduating from Cambridge. During the war he had been employed by the Ministry of Agriculture and became closely involved with the importation of tractors and machinery to Britain from America. After the war he became bursar at Christchurch, Oxford and had responsibility for their 30,000 acres of agricultural estates throughout the country. In 1952 he was appointed Rector of King's College and moved back north with his American wife. Bosanquet knew that there was an urgent need for change within the Department of Agriculture at Newcastle. Under Wheldon, it had become far too parochial and introspective, with something like 80 per cent of the staff having themselves graduated at King's. He needed someone with courage and determination and with a reputation of making things happen.

Having identified the man, he needed to spark his interest. He therefore invited Mac and Hilary up to Rock for the weekend and showed him the College, the farms and perhaps just as importantly, the beautiful countryside of Northumberland. No doubt Bosanquet's wife, who herself had been less than enthusiastic to move to the north from Oxford, had the job of selling the area to Hilary. Mac could see for himself the enormous scope and challenge of the place and what was more, he recognised in Bosanquet, someone with whom he could work, and from whom he would receive support and encouragement.

Undoubtedly one of the key attractions of the job for Mac would have been the fact that the Rector had managed to secure the position of Dean of Agriculture as a post outside the Durham University Act. This meant that it was a permanent appointment and thus enabled the holder to make fundamental changes both to the curricula as well as to the strategy of the two farms, without the fear that unpopular moves might adversely affect his tenure. Finally they both knew that a great deal of money was needed to modernise the farms and rejuvenate research work which would attract post graduate students. As Rector, Bosanquet was well placed to give Mac the assurance that these resources would, in due course, materialise.

Clive Dalton's description of Wheldon seems pretty accurate. Fred Blackburn had come to King's as an undergraduate in 1947 and after gaining a degree in Agricultural Botany, he joined the staff as a Demonstrator and remained on the staff as a lecturer until he finally retired in the 1980s. He tells the tale of being summoned to Wheldon's office one day and told to do microbiology, which he then proceeded to do. No one dared to argue. Wheldon, himself a former student in the days of Armstrong College, was a local farmer and successful Jersey breeder. Apart from his sartorial elegance, Fred remembers that he also for a time used to arrive at the college in his Rolls Royce. However, after a while even he thought that was a little too ostentatious.

Wheldon had a formidable reputation and he was used to having things done his way. There was no discussion and, rather like a military establishment, the orders came down the hierarchy without any queries going in the opposite direction. His two key staff were Jim Hall, senior lecturer, and Sandy Main, although there were a couple of Professors as well. Hall was certainly assumed to be the front runner for the vacancy when it unexpectedly arose and some even considered Sandy Main to be a good outside bet. There was therefore dismay and amazement amongst their supporters when the decision was announced.

It wasn't long before there were repercussions. Jim Hall left to take over as Principal at Newton Rigg, the Cumbrian College of Agriculture. Very much a leading light in northern farming circles, he clearly felt that to play second string to Cooper was not going to work. Whilst this 'half-cock' colonial might have impressed the gentry down south, the men in the north were going to need a great deal of convincing that he knew what he was talking about. Sandy Main as Senior Tutor, however, remained at Newcastle for the rest of his career. Again he was a local man and well respected by the farming community. He was an able administrator and a political 'fixer', as well as being an enormous help to Mac in his first few years.

There were other academics in the Faculty when Mac arrived, including Professor Cecil Pawson, another King's graduate, who had responsibility for the scientific work at Cockle Park. Pawson had started his career as Farm Recorder in 1916 and worked his way up from lecturer to Professor over the following thirty-six years. He was engaged in writing the history of Cockle Park and seemed not to mind being sidelined to finish that job, thus making way for Mac to take direct responsibility for the research farm.

Someone else who was to play an important role at Newcastle for many years, was Jim Merridew. A large, jovial and rumbustious character, Jim had taken a degree at Aberystwyth and gone to Newcastle as a postgraduate student, studying for an MSc in agricultural engineering. In 1951, Mac was looking for someone who could sort out the farm machinery problems at Wye. Merridew's name came up in conversation, and within a few weeks, he was on the staff as a lecturer, without having to first serve his time as an assistant. This job carried with it the responsibility for all the agricultural and horticultural machinery on the farm as well as the hop growing work. It was invaluable experience and he obviously got on well with his boss. He was both surprised and delighted to be offered a lecturership in farm machinery as well as the post of Director of the farm at Nafferton when Mac moved up to Newcastle.

Nafferton extended to some 730 acres of fairly good land in the Tyne valley about ten miles due west of Newcastle. The farm was part of the Allendale Estates and the tenancy had become vacant shortly after the war, owing to the eviction of the tenant through bad farming practice. Wheldon, who had the ear of the aristocracy at the time, managed to persuade the Trustees that the farm should be let to the College. This was a particularly shrewd move as the University was an ongoing institution, making it virtually impossible for the estate to take the farm back in hand at some future date, should they want to.

When Jim Merridew arrived the farm employed 27 staff. It had a dairy herd of about 100 cows, 200 sheep (a number of which were being sold for cash on the side by the shepherd) with the remainder of the land in arable. The terms of reference for the farm were to run it as a commercial enterprise and to make a profit. In addition, the farm was to be used as a practical demonstration for the students in modern farming techniques and also as the basis for a good deal of case study work. Jim already knew the farm as he had worked there as a post graduate student, so that there were a number of inefficiencies he could eliminate quickly, especially on the labour side. The first to go was the shepherd. However there were severe problems in getting hold of sufficient money to modernise the buildings and to expand the dairy herd.

Jim therefore considered that the only solution was self-help by building up stock numbers as quickly as possible, and then to hold annual dispersal sales at the local market. As a result the realisation of the cattle exceeded considerably their valuation in the books. This rather ingenious procedure successfully confused the College administrators and eventually provided sufficient cash to build a new dairy set-up and to double the herd over the next few years. Mac Cooper needed a little persuasion, in what he suspected might be rather dubious practice. Nevertheless he went along with it, no doubt in his mind justifying the means to achieve the end. Some may have questioned Jim's skills in the lecture theatre, but no one could deny that he had an eye for any opportunity to improve farm profitability.

Another good example of the Merridew initiative concerned the introduction of a battery hen enterprise, only the second unit in Northumberland. An egg merchant, with a 4,000 bird unit in Newcastle, had advertised his business for sale. The location was good for retail sales but he had some fairly substantial logistical problems, not least of which was the difficulty of disposing of tons of poultry manure, especially as the unit was sited on the fifth floor of a building in the centre of the city! After months of haggling, Jim managed to strike a deal with the increasingly desperate producer and bought the whole lot for £2,000. The cages all had to be dismantled and transported down five flights of stairs. Once operational back at Nafferton, the enterprise showed excellent returns for many years until the cages eventually fell to pieces and the activity ceased.

Cockle Park, however, was a very different proposition. Cecil Pawson's book painstakingly describes the history of this Experimental Station from the beginning of the twentieth century until 1956, or effectively, until Mac Cooper began its revitalisation. Very much a scholarly work, Pawson dedicates his book to:

Somerville, Middleton and Gilchrist
whose work, in the realm of scientific agriculture,
has satisfied the most searching of all tests—
that of time

The depressed agricultural industry at the start of this period had resulted in much of the heavy land in the north of the country being ranched, with cattle or sheep grazing extensively over acres of self-sown vegetation of very low value. With the founding of Armstrong College in 1891, its first Professor of Agriculture, William Somerville, became increasingly anxious to undertake on-farm research work, in order to demonstrate the ways in which land productivity might be improved throughout the north east of England.

After years of discussions, the Northumberland County Council acquired Cockle Park and arrangements were agreed whereby experimental work under the auspices of the College could commence. Somerville's aim was to try and find an inexpensive method of improving the production potential of the thousands of acres of pasture in the county. His theory centred on the use of a by-product of the steel making industry, namely basic slag. This material was rich in calcium phosphate, which was in plentiful supply further south in the industrial areas of Teesside. Thus commenced the famous long-term trials in Tree Field and Palace Leas fields. Grass production from the treated fields was measured through increases in the live weight of sheep or the yield of hay. And the results were impressive. The trials went on from that time up to Mac Cooper's arrival, some sixty years later, probably the longest-term experiment ever in the history of grassland farming.

One of Cooper's early controversial decisions was to put the plough into these hallowed acres. He felt that the ground could be better utilised for other work and that times had moved on. Fred Blackburn, ever the enthusiastic botanist, was one of those who at the time considered that this was an act of irresponsibility. He felt so strongly on hearing that the field was to be ploughed that he rushed up to the farm on a snowy Christmas Eve to rescue some samples of turf for future analysis. The ploughing of Tree Field remains a talking point in Cockle Park folklore to this day. Gilchrist followed Somerville as Professor of Agriculture and made what was probably the greater contribution with his work on reseeding grassland, using a mixture of grass and wild white clover. His celebrated Cockle Park seeds mixtures became well known worldwide.

For many generations of students these classical experiments were immortalised in the Cockle Park song, performed with customary gusto on 'Agric. Nights' in the Bun Room, to the tune of 'Hark the Herald Angels Sing'.

> Lime and lime and no manure
> Makes both farm and farmer poor;
> When the wind blows from the east
> Then 'twill rain two days at least.
> Basic slag and wild white clover
> Now renowned the whole world over.
> Cursed be the clumsy clot
> Who's never heard of Treefield Plot
> Hark the herald angels sing
> Basic slag is just the thing.
>
> You must try these famous seeds
> That supply all farming needs.
> Gilchrist's mixtures fine and rare
> Smoked by farmers everywhere!
> Acclamation from the nation
> For our class two weather station
> Those who would their tutors please
> Get genned up on Palace Leas!
> Hark the herald angels hark
> Glory be to Cockle Park.

According to Clive Dalton, one of his more intellectually inclined contemporaries, the late Henry Pickering, added a number of new verses to reflect the changing times. The first went as follows:

> Now the new Dean's here to stay
> Palace Leas may grow no hay.
> Pigs and poultry there you'll find
> Nowt but pork and bacon rind!
> Treefield's reign is now all over,
> Now will grow no more white clover.
> Those who would Prof. Cooper please
> Had better watch their Qs and Ps.
> Hark the herald angels shout!
> Cockle Park turns inside out

It was not until 1947 that King's College formally acquired Cockle Park. Situated about fifteen miles due north of Newcastle, near the town of Morpeth, the 520-acre farm lies 300 feet above sea level. It is within about six miles of the North Sea and with little natural shelter, the conditions are often severe, with grazing not being possible until the end of April. Most of the soil is boulder clay, of only moderate fertility, and typical of large areas of the surrounding countryside. Since the war, the research work continued on much the same lines as it had done in previous years, with the attention turning to ways of effecting an increased degree of self-sufficiency. Feeding trials with root crops and hay using beef cattle and sheep, in buildings ill-equipped for experimental work, kept the staff occupied. There were no dairy cows, pigs or poultry and it was self-evident that the whole place was in desperate need of a major shake up. The problem was that no one had been brave enough to rock the boat. This is what Cooper had been hired to do and he set about it in typical fashion.

Ted Pears started work at Cockle Park in October 1935 at the age of twenty-five. His wage then, in today's currency, was just over 30 pence per week. In 1937 he moved into one of the farm cottages with his wife Edie, and they have been there for the last sixty-three years. Ted was farm foreman when Mac Cooper arrived on the scene and has a vivid recollection of the impact that he made both on the staff and the surrounding farmers, all of whom were waiting with bated breath, to see what would happen next.

At first not much did happen, principally because there were no tools to do the job. There was only one tractor; the rest of the work was with horses and sheer manpower. Within a few months, however, tractors and machinery began to arrive. Plans were finalised to introduce a Jersey herd of dairy cows, with the buildings to go with it. The plough was into Treefield and there was talk of both pig and poultry enterprises due to follow on shortly. Cockle Park was indeed turning inside out.

When the Cooper family moved into Cocklaw, the new house built for them next to the old Pele Tower, Mac was on site most weekends and spent a lot of his time discussing ideas with Ted about how buildings might be constructed, using secondhand materials and farm labour. It was a far cry from the professors of old, who never even talked to the farm staff, let alone invited them to offer their opinions on how things might be done.

When asked how he might summarise Mac Cooper's impact, Ted in his rich Northumbrian accent offered the opinion: 'Never ever, and that's a long time mind, would there be a chap like Prof. Cooper. He was into everything and what he didn't know he had the finest way of extracting it and it was his thereafter.' Ted himself was an institution at Cockle Park and it was to everyone's delight that he was subsequently awarded the British Empire Medal for his years of dedicated service.

But as with all ambitious plans, it took in fact several years before the dairy

Ted and Edie Pears, July 1999.

Cocklaw – the Coopers' first home in Northumberland at Cockle Park.

herd, piggery and poultry enterprises were properly established and worthwhile research projects got under way. Rather more important than bricks and mortar was the need to motivate the staff in the Department and to gain their confidence. Shortly after his arrival and just before Christmas, Mac summoned everyone to a meeting, including the domestic staff. Bill Weeks, a junior lecturer at the time, remembers that Mac came in like a 'dose of salts'. He told them all that in future there would be no fixed hours of work, that his door would always be open and that he expected everyone from professors to caretakers to work towards building Newcastle into the leading Agricultural Faculty in the land. Many were frightened by his blunt approach. He made no concessions to the status quo. Nothing like this had ever happened before. On leaving the meeting, Sandy Main was heard to say that, rather than extending to everyone the 'compliments of the season', it seemed as if they had just received 'the complaints of the season'. But Mac also knew that however well he might generate a new enthusiasm, he also needed fresh blood, preferably a transfusion from New Zealand.

One of the most glaring gaps in the curriculum at Newcastle was the total lack of teaching in statistics and experimental design. Although Mac had little depth of knowledge in either discipline, he realised that something had to be done quickly, as external examiners who were evaluating postgraduate work were highly critical of the shortfall. His first New Zealand import, therefore, specifically to fill this gap, was Rex Patchell. Rex had been at Massey and worked in the Dairy Research Institute for about six years after he graduated. Although rather a serious man, he was an able research worker and a skilled statistician. Mac knew of Patchell's reputation and wrote to Professor Riddet at Massey to see if he could persuade him to take a job at Newcastle, describing it as a golden opportunity to gain overseas experience. The ploy worked and in due course the Patchells set out for England by sea, never having been away from New Zealand before.

When they eventually arrived in Newcastle in mid February, the Cockle Park farm manager met them off the train and handed them a welcoming letter from Mac, expressing regret that, as he was in Egypt, he would see them on his return. No arrangements had been made for their accommodation and they knew no one in this god-forsaken corner of the far north east of England. Hilary, however, came to their rescue, put them up and soon managed to find them a flat not far from the College in Newcastle. But Rex and Pauline Patchell's first impressions were not good, especially as they couldn't stop shivering in the intense cold.

When Patchell started work he was amazed to find a complete lack of understanding of his subject within the Department. Everyone running a research project had strict instructions to check their experimental design with him before proceeding any further. Thus it was that he had a constant stream of people at his door asking for help. The job was clearly going to be a real challenge. As well as teaching basic statistics and population genetics to students, using a battery of hand cranked calculators, Rex Patchell also became involved in the breeding work with sheep and eventually poultry when he moved out to Cockle Park. In association with a newly arrived postgraduate, Maurice Bichard, they developed a programme to demonstrate the possibility of breed improvement through progeny testing with a flock of Clun Forest sheep. Whilst an unusual breed for Northumberland and the butt of uncomplimentary remarks at the local mart, the project continued for many years and was particularly useful for teaching purposes in demonstrating the results of selection and recording within a self-contained flock. Another of Clive Dalton's recollections, when he was helping Mick Givens, the Cockle Park shepherd, to drench the ewes, was Mick's description of their colour. The lambs were blackish brown and got lighter with age. Mick described them as 'a daft sheep from doon sooth' and their colour as 'moosy doon'. When asked for a translation, he went on: 'the colour of a pig's fart in the moonlight'. There clearly existed a healthy gap of opinion between the scientific researcher and the farm worker.

Some would argue that Patchell was the first real scientist to join the staff in the Department and his contribution during the five years he was at Newcastle, before returning to Massey via Australia, was, in statistical terminology, highly significant. Mac Cooper persuaded Bichard to stay on at Newcastle to replace Patchell and continue the sheep work as well as lecture to the students. In fact he remained until after the end of the Cooper era before leaving to take up an appointment with the Pig Improvement Company and gaining an international reputation as a leading animal geneticist.

Another New Zealand import was Mat Sanders as farm manager at Cockle Park. Mat had been a student of Cooper's at Massey and one of the party who had given him such a fond farewell when he left for Britain in 1947. Mat was to replace Paul Brennan, who had moved on to pastures new. Brennan had been a 'collar and tie' man, whereas Mat Sanders was a typical brash 'kiwi'. He didn't understand that there might be protocols and red tape involved before things could get done; he just did it and asked questions afterwards. Mat brought with him a refreshing blast of energy to Cockle Park and whilst it took a little while for him to understand the natives and for them to understand him, he soon earned their respect. Under the old regime there would have been endless meetings, involving layers of senior staff, before any decisions would emerge. Furthermore it released Mac from the hands-on control he had assumed when he had arrived at Cockle Park. He now had one of his own breed to get on with the job and as long as he was kept fully informed, then he was happy. The two of them worked well together, although it has to be said that Mat sometimes had difficulty in accommodating the inconvenience of the research worker's requirements, when what he really wanted to do was get on with the practical farming.

One of these researchers, although not employed by the College, was David

Rivett. The National Institute of Agricultural Botany had national field trial sites throughout the country, for the purposes of evaluating new varieties of crops prior to publishing the results for use in commercial farming. Rivett was their man at the Cockle Park site. The work required detailed recording and analysis of large numbers of replicated plots. It was painstaking work and required, certainly in Rivett's view, instant availability of both labour and machinery at short notice, especially if the weather was inclement. Unfortunately if the weather was difficult for the experimental work, it was also difficult for the farm; thus priorities were a constant source of argument. Rivett had many qualities, but patience and understanding were not amongst them. With an unruly crop of red hair and a face to match, his temperature rose quickly when things didn't go well. Many were the times that a slanging match between the farm manager and the NIAB Trials Officer reverberated round the farmyard. But most of the time it was fairly good-natured leg pulling. Rivett was also the unwitting victim of many practical jokes, usually initiated by Mat Sanders, but often in close association with the Dean. On one occasion he was asked to go and collect Mac from the Central station in Newcastle about fifteen miles away at 5 am. It was not until he arrived that he found that there were no trains due in from London until late morning. It also gradually dawned on him that it happened to be 1 April.

In addition to all the demands for new buildings to enhance research facilities, there was also an urgent need to provide accommodation for the increasing

Clive Delton, Mat and Anne Sanders, April 1999, NZ.

number of postgraduate students engaged in experimental work at Cockle Park. This came in the form of renovations to the old Pele Tower. These were, however, exceedingly rudimentary and the infamous north-east gales continued to blow through the gaps between the window frames and the old stonework. The inevitable solution was that many of the incumbents preferred to spend most of their winter evenings keeping warm in the Oak Inn, an isolated public house at nearby Causey Park. Eventually conditions did get better, but for many of the early pioneers in the Cooper era, everyone had to make do and mend. This also applied to the research facilities. Whilst a field laboratory had already been built, postgraduate students were expected to fence their own field plots, build digestibility crates for sheep and construct experimental silos in which to conserve grass. This was the atmosphere at Cockle Park in the early sixties. New people were arriving, everyone mucked in and there was a general air of enthusiasm, hard work and good fun. This period saw the gradual completion of the buildings to house the pigs, dairy herd and beef cattle.

Pigs had always had a special place in Mac Cooper's heart and he was keen to see Cockle Park develop an up to date unit that could provide 'state of the art' research facilities for postgraduate students. Prior to building the unit, Mac had agreed to allow Bill Smith, who had recently returned from Massey, having completed his MSc in silage research, to come to Cockle Park to work for a PhD. This was in response to a personal recommendation from Professor Riddett and although Bill had never worked with pigs, Mac already had him in mind as a likely candidate to head up the pig research programme.

Bill, a Scotsman, arrived with his New Zealand wife Margaret to find, somewhat to his surprise, that his first job was to help build the unit. With his highly developed sense of dedication, he set about this with gusto and at the same time worked with Mac to plan the experimental work to be undertaken. The Smiths lived in nearby Morpeth and had no car. This meant that Bill often had a three mile walk from the bus stop on the main road before he got to the farm, especially at weekends when there were fewer colleagues about to give him a lift. It also meant a lonely life for Margaret, but as usual Hilary was on hand with support to help her acclimatise to the area and make new friends.

Mac had already established the Co-ordinated Large White Breeding Project based upon performance testing which involved eight herds belonging to agricultural teaching institutions in Northern England and Scotland. This project was followed by a wide range of studies utilising the new unit, including breed and carcass evaluation as well as investigations into husbandry systems. Bill Smith took a broad and practical view of research and the place it had to play in helping producers improve efficiency. He often involved nutritionists, microbiologists and economists, thus taking a whole industry view rather than being just concerned with genetic or nutritional problems. Furthermore he could talk in terms which farmers could easily understand. He was just the man for the job.

Bill and Margaret Smith left Cockle Park to go back to Massey after the Coopers moved to Spain in the seventies. Tragically, however, Bill's highly successful career was cut short when he died prematurely from a heart attack soon after their return to New Zealand.

The new diary unit enabled work to commence on feeding trials to improve the proportion of milk that could be produced from grazed grass and silage. With hindsight, this activity may well have made more impact if the herd had been Friesian rather than Jersey. Whilst at the time there seemed to be potential for the Channel Island breed in terms of higher milk prices and increased stocking rate, subsequent events proved that the black and white animal was to become the preferred route to profitable milk production. Mac's preference for the Jersey stemmed from his experience in New Zealand. He seemed to have made up his mind about the choice for Cockle Park some time previously, without perhaps anticipating the rapid expansion of the black and white breed in both countries. In Britain especially, the beef industry relied on the supply of most of its calves direct from the dairy herd and as the Friesian gained in popularity, the demand for bull calves and beef crosses became a significant source of supplementary income for the dairy farmer. Channel Island cows had a much lower value in terms of both the calf and the cull cow and these milk producers largely missed out. The herd at Cockle Park did eventually change to Friesian, but rather late in the day.

Although the rapid expansion of the Friesian significantly lifted national milk yields per cow, there evolved a heated debate in breeding circles as to whether the right strategy towards increased profitability lay in selecting for more extreme milk production, or in persevering with the dual-purpose policy of more modest milk yield and a good calf value. The first option would have meant the importation of Holsteins from North America. With the health safeguards then in place to avoid foot and mouth disease, such a policy however failed to gain Ministry support. The result was that the black and white cow population remained predominantly Friesian and the genetic improvement in milk production in the UK virtually came to a standstill. Mac Cooper did not become involved in this national debate which rumbled on until the mid eighties when eventually it dawned on everyone that the rest of the world had opted for the Holstein and that it was to start catching up. Had he decided to join the fray, he would have found plenty of vested interest and bureaucratic red tape to get his teeth into, but with a Jersey herd at Cockle Park, his credibility was rather limited.

Whilst the dairy industry lost ground in genetic improvement as a result of this long running debate, the beef sector flourished with the Friesian influence and the subsequent introduction of continental beef bulls for use in cross breeding. The Milk Marketing Board, in response to their producer's demands for better information as to which were the best beef bulls to use in AI, set up a large beef progeny test station at Warren Farm in Wiltshire. At the same time many of the Research Institutes and University farms started research programmes geared to evaluate the new breeds and to find systems which would best exploit their potential. Cockle Park was no exception. With the new beef building coming on stream, the earlier work carried out by Ian Mason at the Edinburgh Institute of Animal Genetics and by another New Zealander, Viv Vial at Cockle Park, on evaluating Friesian bulls for growth and carcass traits in their male offspring, was followed by more applied systems work. In fact the beef work at Cockle Park during the sixties covered a lot of ground, including rapid rearing of bulls on high cereal diets, eighteen month finishing on silage, and fattening cross-bred heifers after they had produced one calf.

A systems approach was applied in the sheep sector as well. Bob Thomas and later Murray Black were deeply involved in developing methods to control parasitic worm infestation in lambs through clean grazing techniques. This was followed by investigative work into the incidence of metabolic diseases, particularly grass staggers, illustrating the vital role which magnesium played on pastures which had received heavy dressings of nitrogen and potash.

The topics for these research projects did not arise by accident. Mac Cooper recognised that they addressed problems that farmers were coming up against as the new technology evolved. He first made sure that he had people with the dedication and enthusiasm to tackle the job and then told everyone who cared to listen about the work in progress. He urged farmers to come and see for themselves. As a direct result, the number of visitors to the farm escalated. Mac conducted these farm tours whenever he could and there was nothing he enjoyed more than showing groups of farmers round, challenging them to interpret the results with regard to application on their own farms. Whilst inevitably he took centre stage, he was careful to insist that the researcher explained his own work. He saw this as an opportunity for young people to take the limelight and in so doing gain in self-confidence. Should they come under too much pressure or get out of their depth, he came to the rescue. He would talk quietly to them afterwards to point out areas where they might, on future occasions, make improvements in their presentations. This was a sure way of keeping research workers' feet firmly on the ground. It wasn't the project itself that was all that mattered; it was the skill of getting the message across that carried just as much importance. If they could talk convincingly to a group of Northumbrian farmers, then in Mac's view they must be heading in the right direction.

Most of the visitors were farming groups or overseas research workers, although occasionally a major event, such as the summer visit of the British Grassland Society in Mac's presidential year, brought out all the stops. Such an occasion caused a great deal of preparation and tidying up. Mac could be quite fastidious and woe-betide anyone careless enough to leave litter lying about or cattle yards insufficiently bedded down with straw. Both farm staff and research workers were careful to hide all evidence of their carelessness well before the party arrived.

If Newcastle was to further its reputation there had to be more than good research and teaching facilities. There had to be a national figurehead who attracted attention. Mac relished this challenge and recognised the need to keep in the public eye. It wasn't too difficult in that he continued to receive frequent invitations to speak at conferences and farmer meetings as well as participate in radio and television broadcasts. Soon after his arrival at Newcastle he had returned to Oxford to address the ninth Farming Conference. He was one of three speakers in a session entitled 'Agricultural Production and Marketing Problems Today'. The economist on the platform issued dire warnings that farmers were becoming far too complacent and that Government was making noise about reducing the £200 million subsidy into the industry. Whilst the politicians heard the pleas from farmers that they would prefer to see tariffs imposed on imports rather than subsidies on product prices, Government were not prepared to risk retaliatory measures from the same countries who were buying British manufactured goods. He went on to urge the audience that the era of food at any price was over and that farmers must look at ways to reduce production costs if they wanted to increase profits. The second speaker was an MP and farmer from Cambridgeshire, who advocated the need for continued stability through an extension of the guaranteed price mechanism. He waxed eloquent about the success of the Milk Marketing Board and whilst there were substantial difficulties in copying this approach in other sectors of the industry, he saw great potential for greater co-operation in marketing for the benefit of producers.

Mac Cooper on this occasion was perhaps a little less controversial than might have been expected. Certainly on his old hobbyhorse of subsidies, he was surprisingly restrained. On the cereal side he was complimentary about British farmers and their ability to produce grain cheaply and efficiently, especially on the larger, well-equipped units. Within the livestock sector, however, he was less generous. A fifteen-year period of fairly good prices had obviated the need to look critically at production methods. Pigs came in for sharp criticism as quality had declined at the expense of quantity. Housing conditions, animal welfare and poor selection of breeding stock had created an unwieldy edifice which the Danes were rapidly undermining. Furthermore there were far too many hangers-on taking a cut out of the producer's margins. These extended from over expensive feed compounders to unscrupulous middlemen in the buying chain. The sooner farmers woke up to this the better. He then turned to what he described as the 'sheet anchor' of British farming, the dairy industry, in which he felt there was just as much scope for improvement as there was in pigs. The major difference was, however, the competition within the industry, rather than as in the case of pigs, the competition from potential importers.

Because dairy farmers now received virtually the same pooled price for their milk, those who were to be successful would have to squeeze the small and inefficient producer out of business, through increasing herd output to spread overheads and reducing feeding costs by improving grassland management. At the Oxford Conference, such an assertion would have been unlikely to cause much dissent, as the audience would contain hardly any small farmers.

The discussion following the papers at Oxford is often more enjoyable than the contributions themselves. On this occasion there were many speakers, including a Colonel R. Henriques, who opined that whilst he did not want to appear ungrateful, he felt the first two papers 'have about as much relation to the realities of production and marketing during the last three months, as the stories I was taught as a boy, about the baby being picked off the gooseberry bush and brought through the window by a stork.' Most contributors in the audience felt that the real villains of the piece were the large fertiliser, machinery and feed compounding firms, who were taking unreasonable profits through over pricing their products. There seemed little evidence, at least from reference to the published proceedings, that the audience felt the future lay in self-help. They appeared to prefer the rather easier task of continuing to identify scapegoats upon whom to heap blame for their difficulties.

Mike Soper, who edited the proceedings of the conference and many others which were to come in future years, dryly observes: 'Some things might almost be said to apply today; but then everything seems to go round in circles in farming, even if it takes fifty years or so to come back in another guise.'

As well as being a gifted speaker, Mac Cooper possessed a talent for the written word. He published a number of books, which for the most part focussed attention on livestock production. One of his first, and arguably his most thought provoking was *Competitive Farming*, commissioned by Crosby and Lockwood in 1956. The recently published Government White Paper on the Annual Review of Agriculture included the stated objective: to achieve a steady improvement in the competitive position of the industry. In the publisher's view, they felt someone with vision and knowledge should spell out how this might be done. Their request was prompted by the fact that Cooper had something of a reputation for plain speaking and he had already stimulated considerable discussion on a wide range of topics. These needed pulling together in an authoritative way.

Mac dedicated his book to Professor Riddet at Massey, for whom he retained the highest regard, and in the Preface, he states: 'This book is in no sense a blueprint for agricultural development, rather it is intended that it should make people think. The salvation of British farming lies not in the preaching of professors but in a thoughtful and progressive attitude to the industry by the general body of farmers and by the men who administer it.' The book received wide acclaim and inside the front cover of the copy which he inscribed and gave to Hilary, there is attached a personal letter.

From the Minister of Agriculture, Fisheries and Food
Whitehall Place, London, S.W.1

15 October 1956
Dear Professor Cooper

I have just finished reading 'Competitive Farming' and should like to tell you how stimulating I have found it. You won't expect me to say I agree with all you say, but you have certainly given us a great deal of food for thought.

I see that you warn us about this in your preface to the book where you say that your aim was to set out views on a number of problems, 'stated with the express purpose of making people think'. I am sure that you are succeeding admirably in doing this.

Yours sincerely,

D. Heathcoat-Amery.

To those who had listened to or read about Mac Cooper's opinions since he had arrived in England some nine years previously, much of the contents of this book would have been predictable. But in a composite form, the comprehensive nature of his analysis was impressive. Moreover, as well as stating what he considered to be the main problems, he offered practical proposals as to how progress might be made towards their solution.

His antagonism towards subsidies was well known and despite the unpopularity that their abolition would generate throughout the farming industry, he insisted that they must be progressively removed. Not only were they an unacceptable burden on the taxpayer, they killed initiative and kept inefficient farmers in business. Furthermore 'they hold the carrot just a little too close to the nose for the good of farming and they tend to bring opprobrium to the industry.'

He recognised, of course, that British agriculture, if it were to become internationally competitive, would continue to need substantial funding from central Government. But this must be channelled, not into subsidies to support prices, but into capital injection to modernise outdated farms and practices. Machinery must replace labour so that overall costs of production could be reduced. Furthermore the State should introduce a land settlement policy, similar to the one which operated in New Zealand, and designed to give young farmers with initiative and ability, but little capital, the opportunity to get their feet on the first rung of the farming ladder. This would introduce new blood into an industry which badly needed it.

A large section of the book dealt with the high cost structure of the livestock industry, especially pigs and dairy cows. In the pig sector, he was scathing in criticism of the guaranteed price incentives, which had been introduced to encourage rapid expansion and to help offset the post war meat shortage. He repeated the accusation he made at the Oxford Farming Conference in the previous year, in that this had been an example of short-term political expediency of the very worst kind. In the book, however, he went on to propose the setting up of a Development Board for the pig industry, which would take responsibility for progeny testing and pig recording. This would operate much more effectively if it were outside the Ministry of Agriculture, who he felt were not up to the job. There was tremendous potential to increase farrowings per year, litter size and feed conversion efficiency and farmers needed specialist advice, based on reliable information, to help them achieve these gains. This was not a job for the overworked District Advisor. Specialists were needed who knew the answers and in whom pig producers would have confidence. If an example were required, then one needed to look no further than the subsidiary activities which the Milk Marketing Board were undertaking on behalf of their dairy farmers.

A number of chapters dealt with grassland and the potential to increase production as well as to reduce overall feed costs. Comparisons between the extremes of technical and financial performance, published by the Survey of National Milk Cost Investigation, vividly illustrated the vast range between the best and the worst practitioners. The means by which farmers could dramatically improve their profits were there for all to see, but real incentive was needed to effect change. Over generous prices, however, continued to keep inefficient farmers in business, when they should be forced out to make room for those who were prepared to move with the times.

Another familiar theme was livestock breeding and the vested interest of some of the more misguided pedigree breeders and their societies who were fighting a rearguard action against the application of genetic evaluation of breeding animals through progeny testing. The Ministry of Agriculture also came in for its share of ridicule over bull and boar licensing.

It is vaguely realised in high places that licensing on its present basis is not doing enough, so at intervals committees are set up to investigate its efficacy. Unfortunately, the committees are weighted by people who either don't know or whose livelihood is concerned with the administration of licensing. If the generals had had the decision as to the size of the standing army after the war, very few of us would have got out of uniform for people don't destroy the things that make them important. So it is with bull licensing. The minorities on the committees who have some knowledge of genetics or some comprehension of the futility of licensing, may make their protests but unavailingly.

In a chapter entitled 'Knowledge is Golden', Mac opens up a new area about which he has strong views: the need for an effective advisory service. The National Agricultural Advisory Service (NAAS) had been established for about ten years, but in his view, it had a long way to go before it would gain the full confidence of farmers. In fact, he suspected that technical feed and fertiliser company representatives probably had greater influence. This, he insisted, should not be taken as a criticism of the individual, many of whom did an excellent job, in spite of the nine to five armchair administrators who forever queried their expenses. The service, however, had singularly failed to penetrate the poorer half of the farming population, the very place where it was most needed. There prevailed a deep suspicion of Ministry legislation, inspection and interference, probably a legacy of the old War Agricultural Committee days. Try as the individual officers might, this reputation could not be changed.

The solution Mac put forward was a complete break away from MAFF and the positioning of the Advisory Service outside all the administrative tasks of licensing, checking on subsidiary applications and inspection of premises. There must not be any civil service element or agency responsibility to Whitehall. The advisory staff must have only one function: the welfare of their clients. He was less specific, however, with regard to the organisation that might take over the responsibility for such a challenge, alluding to some sort of farmer controlled representation. This may well have been with tongue in cheek, as farmer politicians were just as capable of getting hold of the wrong end of the stick as were their contemporaries in Westminster or Whitehall. One of the more obvious options may have been to position the Advisory Services within the Agricultural Departments of the universities, as was the case in Scotland. Mac, however, did not enthuse about this approach, mainly because he felt that professors were already over-burdened with administrative detail. He complained, 'Committees of various kinds are the bane of the average professor's life and he indeed is fortunate when he comes to the stage of eminence that he has two meetings on the same day so that he can cheerfully send apologies of absence to both.'

Cooper's book attracted favourable reviews in the national press. Clifford Selly, a respected agricultural correspondent of the time, wrote:

Cheap imported food is the traditional bogey of the British farmer, and now that imports are rising and prices falling at the farm gate, if not in the shops, farmers are inevitably on the defensive. Any attempt to secure a revision of the present subsidy system, now costing the Exchequer about £200 million a year, is greeted with cries of 'Why pick on us?' Indignantly and not unreasonably farmers point to the tariffs, price rings, monopolies and other devices for exploiting consumers which are enjoyed by urban industries. Surely, it is suggested, farmers are entitled to some protection, and without subsidies they would be lost.

It is this fear that reduced subsidies would spell farming disaster, which has stifled objective discussion of agricultural policy in recent years. But here at last [review of *Competitive Farming*] is a farming authority who is not on the defensive and without any of the usual axes to grind, a courageous man who insists that the farmer, let alone the taxpayer, is not getting the desired benefit from the present reluctant State dole.

Lord Bledisloe also felt inclined to add his considerable weight to the ensuing debate with a rather long-winded letter to *The Times*. He anticipated that the book would no doubt 'evoke vehement criticism among bounty-fed British farmers, as it repudiates agricultural subsidies.' He concluded his letter by saying: 'Professor Cooper emphasises the enormous potentialities of properly managed grass and leguminous herbage in the greater production of meat and milk. My own experience during the last decade on good land in this west country (if well drained) is that an increase in output of at least 33 per cent may be confidently anticipated from the adoption of methods such as Professor Cooper advocates.'

Competitive Farming was not Mac Cooper's first book. When a young lecturer in New Zealand, he had been one of the co-authors supporting McMeekan, who published *Principles of Animal Production* in 1943. This was largely based on the original work carried out in Cambridge by Hammond and adapted to the conditions prevailing in New Zealand. The authors were determined to present this mainly scientific material in terms which farmers could follow. Furthermore, they were sufficiently dedicated to contribute all the royalties from the sale of the book to 'the assistance of promising students of animal production'.

Mac enjoyed writing. During the twenty-five years he spent in agricultural education in the UK, he also produced books on sheep farming and beef production. These were written in conjunction with research colleagues at Newcastle: Bob Thomas in the former case and Malcolm Willis in the latter. Not only did these works focus on the practicalities of making money out of grazing livestock, but they were also a means by which prominent research workers at Newcastle could highlight their findings. *Grass Farming*, probably the best known of Mac's books, was first published in 1962 and ran to five editions over the following twenty years. Dai Morris, a former postgraduate student at Newcastle, a future son-in-law and later, the first Principal of the Welsh Agricultural College, joined Mac as co-author for the third edition. In the preface to the first edition of *Grass Farming*, Mac makes the point that he had written the book more for farmers than students. He goes on: 'I sincerely hope that it will hold their interest, for I realise that there is no soporific more powerful than the printed word for anyone who has spent a long day in the open air.' Whilst Mac maintained that these books were not intended for those who wished to pass examinations, the fact remains that they became standard texts throughout the country for generations of aspiring agricultural students.

Mac was also keen to increase his international experience as this not only stimulated new thinking, but it helped to project his work at Newcastle to a wider audience of potential students from overseas. He had already completed a number of external examining roles in Africa; however he was also very keen to engineer a trip home to New Zealand, preferably at someone else's expense.

This came when he was invited to speak at the Seventh International Grasslands Congress at Massey. Sadly, Hilary felt unable to go with him, partly because of the cost, but mainly because the girls were too young to be left for such a long period. Since he left New Zealand, Mac had acquired a formidable reputation in both countries and an article in one of the newspapers, welcoming him home, was prefaced by the following paragraph:

> At a luncheon of agricultural journalists in London, Percy Cudlip, celebrated London editor, laughingly termed Professor Cooper, 'the most controversial so-and-so in English agriculture.' Professor Cooper regarded the comment as the biggest compliment paid him since a spectator at an Army Rugby match in Italy expressed an opinion in even more unparliamentary language.

At the congress he spoke with his usual frankness, maintaining that his major contribution in Britain had been more in terms of presenting a fresh viewpoint, rather than anything particularly academic. Such a tactic was bound to be controversial. He also felt that farmers in New Zealand were progressing faster than their counterparts in the UK. There was an enormous amount of tradition and many of the farms in England, because they were so badly laid out with labour-intensive buildings, were simply not capable of taking advantage of modern methods. Whilst there had been a 60 per cent increase in production since the war, there remained a vast range in farmer competence and farm profitability. Most of the increase in productivity had come from the top 15 per cent of farms, about which he pronounced, 'this minority are excellent farmers; the majority, however, are mediocre, while another minority are beyond description.'

Mac enjoyed an extended visit of over three months to his home country, spending time with his family in Havelock North and surrounding Hastings, also visiting the South Island, which he felt had many similarities to Northumberland. He spoke at scores of meetings, visited research stations and generally re-charged his batteries. Surprisingly he also decided to return by ship, which added considerably to the length of his absence. In a letter to Thelma before he left for New Zealand, he remarks: 'In many ways I should fly back to save time, but I am longing for the complete rest of the month at sea, as I have not had a real holiday since I have been here. I felt I couldn't face the job here within 4-5 days of leaving you, which would have been the prospect, if I had flown both ways.' Sally, his niece and Thelma's eldest daughter, accompanied Mac on the voyage to England, where she was to spend a year visiting the country.

Back at Newcastle, there were murmurings about the 'Invisible Dean'. Although by now he had gathered a significant group of new staff to help push through the changes he considered necessary, there remained a few of the old guard, watching for any signs of a bursting bubble. Mac was in some dilemma at this period in his career. As Dean of the Faculty, he was loath to delegate. He wanted to ensure that he retained complete control. At the same time, he felt it his duty to continue to crusade and galvanise British agriculture. The former ambition required him to be easily accessible; the latter, by definition, meant his being away from Newcastle on frequent occasions and in the case of the visit to New Zealand, for long periods. Few in the Department, including Sandy Main, relished the thought of a shouting match with the Dean, especially if action was taken in his absence with which he might not agree. The newer breed, however, felt more confident and they forged ahead with their work, sure in the knowledge that they would have Mac's support even if they made an error of judgement. They knew that it was essential that he remained in the public eye and took every opportunity to broadcast the news of the exciting work going on at Newcastle. But the older members of staff and particularly those involved in the more fringe academic activities experienced some build up of frustration because they were unable to discuss their problems with the boss. Such an atmosphere encouraged resentment in some of the quieter corners of the Senior Common Room. There is no evidence, at this point, that Mac was fazed by discontent in the ranks. His character was such that if he felt there was a need to attend more closely to any one aspect of the job, he would do this in addition, and not in substitution for the other things he felt important. Not even his antagonists could deny his commitment or dedication to the cause.

Shortly after his return from New Zealand, the Principal's job at Massey became vacant when Professor Peren retired. No doubt detailed discussions had already taken place about Mac's availability during his recent trip. Everyone would recognise that if he were to return to New Zealand, this was the one post which would tempt him. In the event, he decided against it and Alan Stewart, who at that time was still working for the Milk Marketing Board, got the offer and accepted. It sounds as if it had been a difficult decision for Mac, judging from a Christmas letter he wrote to Thelma. He omits, however, to mention Hilary's opinion, which is unlikely to have been wholly enthusiastic, following her previous experience of life in New Zealand.

Alan will make a first class Principal of Massey. He has shown in his work over here that he has imagination, drive and a capacity to organise. I think he will provide the shot in the arm which Massey needs. My own feelings about the post have been very mixed. As soon as I got back I realised that I could not drop my present commitments. It would have been like trying to let a house, which had neither roof nor weatherboarding. Everything I am attempting is only half done and I would have left here with the feeling that I had run out on the job. It would have meant, too, a splitting up of the family, as Barbara is committed to her medical course for four more years. I would have had to sacrifice a lot of super-annuation too, as my present policy is not transferable. It was very hard when I came to the point of making a final decision, because I always had the feeling that my climax ambition was to be head of Massey. When it came to the point I was actually higher than Massey in my status, though status has been the last consideration in the decision I made.

This decision not to pursue the Massey job was a turning point. Perhaps the letter to Thelma was his way of explaining that his commitment to Newcastle was more important than his personal preference. But he knew that he could not go back to New Zealand and at the same time keep his family together. Times had moved on. He had achieved celebrity status and whilst he preferred not to admit it, he enjoyed his hard earned reputation for stirring up controversy. To return to New Zealand and his adoring relations was unrealistic. It was the price he knew he had to pay. It was clear that Mac's contribution was going to be in terms of furthering the interests of UK rather than New Zealand agriculture.

Life back in the fast lane at Newcastle remained as hectic as ever. His health as he approached fifty was not particularly good and he suffered a particularly nasty bout of pneumonia, no doubt aggravated by his continued addiction to tobacco. Both he and Hilary continued to prefer to 'roll their own' cigarettes, perhaps in the belief that this saved money. Mac also smoked a 'Sherlock Holmes' type of pipe, so his skill in burning holes in his clothes and everybody else's carpets continued unabated. After a brief period of convalescence in the Channel Isles, however, it was back to the unrelenting pressures of the job, without any heed to the toll that his life-style was exacting.

He had made a tremendous initial impact at Newcastle. Now he was determined to build on these foundations and create, as he put it in an article about Cockle Park in the *Journal of the King's College Agricultural Society*, 'the most highly regarded field centre for agricultural research and teaching in the Commonwealth'.

9

NEWCASTLE: 'ONE MAN PLAYS MANY PARTS'

Mac Cooper was approaching the peak of his reputation as he passed his fiftieth birthday. There was still a tremendous amount to do, but the momentum that he had initiated was, by then, in full flow. The early sixties saw major changes at King's College: first the separation from Durham, as Newcastle achieved full University status, and second the building of a new seven-storey block for the Faculty of Agriculture.

But perhaps even more significant than both of these events for Mac, was the publication of a recent Government report into the state of higher education in Agriculture. This report concluded that there ought to be some rationalisation within the Agricultural Faculties of the universities. The publication of empirical comparisons between the different establishments made it clear that Newcastle must produce better student to staff ratios or there could be a threat of closure. Some notable names including the Departments at Cambridge and Leeds had already been consigned to history.

Although this turn of events was alien to Mac's nature, it is very much to his credit that he recognised that this was a problem that was not going to go away. Regardless of how excellent the work or the teaching in the Faculty, judgement in future would be made by external, faceless, accountants and administrators. The bottom line was that either staff numbers had to reduce, or the number of students had to increase.

The major limiting factor determining student numbers within the existing general agricultural course was shortage of laboratory facilities. But even if these were expanded, at significant capital cost, it would not provide the sort of ratios of staff to students that would ensure safety in the longer term. The answer must lie in diversifying the activities within the Faculty, away from general agriculture and into the relatively new fields of rural economics, environmental science and marketing. Laboratory facilities would not be needed for these students. Thus began a change in direction which over the next thirty-five years would eventually transform the traditional School of Agriculture into a

Faculty of Agricultural, Biological and Marine Sciences.

E.M. Bettenson, in his book, *The University of Newcastle upon Tyne, A Historical Introduction 1834-1971*, documents the long running debate with Durham concerning separation of King's College into a University in its own right. As early as the beginning of the twentieth century there had been a 'certain amount of agitation in favour of a separate University of Newcastle'. However, the arguments had ebbed and flowed over the years without a real will on either side to grasp the nettle. As time progressed it became obvious that change was inevitable, especially as an increasing number of Durham graduates, whose courses were exclusively conducted in Newcastle, had never even visited Durham. Bettenson described, 'It seemed to be not so much a case of the tail wagging the dog as of pretending that the tail was the dog.'

Bosanquet, as Rector of King's, would have played a key role in the negotiations that were eventually finalised in 1963, with the handing over of responsibilities. The formal ceremony at the beginning of August was, however, a low-key affair and thereafter, apart from the obvious changes to the signs and the notepaper, life continued as before. As expected, Bosanquet became the first Vice-Chancellor of the new University of Newcastle, thus guaranteeing a continuation of the interest and influence at the very top of the administration for the Faculty of Agriculture. The Duke of Northumberland, who also had a considerable interest in matters agricultural, was installed as Chancellor.

In a note published in the *Journal* of the newly named '*University of Newcastle upon Tyne Agricultural Society*', Mac Cooper refers to Bosanquet's forecast at the time of his appointment: that within a decade, there was a very good prospect that a new building would materialise. It seemed at one time during this period that the priority for this had slipped to the bottom of the list, which had prompted Mac to plead that the Dean of Agriculture 'should be provided with a bicycle so that he could keep in contact with his spreading dominions.' Within the new university status, Mac found himself appointed to the Senate, effectively the governing body, where he was able to re-emphasise the urgency for a new building. No doubt with the Vice-Chancellor's support, he argued that this was essential if the Faculty was to expand sufficiently to meet the demands of the recently published review of agricultural education. Otherwise, it might well not survive in the longer term.

Before the move, the Faculty was housed in six separate buildings, some of which were converted houses, containing small departments such as Economics and Agricultural Zoology. The main block, situated in the original college quadrangle, remained the focal point, but the inconvenience and isolation of the outlying units did little to encourage an integrated approach to teaching and research. Whilst few would enthuse about the rather grim appearance of the new half million pound 'concrete office block', the advantage of having everyone under one roof was self-evident. Cooper, in his report to the Agricultural Society, added:

> Without a doubt the unification of the Faculty is the greatest single advantage of the new building. One appreciates this point when watching the groups that take morning coffee together in the Common Room. Invariably these are heterogeneous in composition with chemists, economists and engineers sharing discussions. A pleasing feature is that postgraduate students are able to join in these coffee-time discussions and feel they are an integral part of the School.

There were other important advantages associated with the new building. A re-equipment grant enabled a badly needed overhaul of the laboratories and also funded the installation of controlled growth rooms for experimental work on both plants and animals. As well as a seminar library and excellent lecture room facilities, Mac reported that 'there is a fine view from the roof, which is now the highest vantage point in the University. It is good for morale to be able to look down on the rest of the University.'

Mac Cooper at 49.

151

The first significant growth area in the move towards expansion in student numbers occurred as a result of the setting up of a marketing department. Mac's own interests in the late fifties had been drawn to the persistent inefficiencies in the marketing of meat in Britain and he saw this as a topic where he could stir up controversy both nationally and locally within the farming community in the North-east. But he needed a platform.

Bill Weeks graduated at Newcastle during the war and returned to King's as a junior lecturer in the Economics Department in 1946. He was one of the 'inbred establishment' when Cooper burst in on the scene some eight years later. Unlike many of his colleagues, however, Bill saw this intrusion as just the sort of tonic required to jolt the place out of its complacency.

Weeks was a colourful character, full of enthusiasm and a natural performer in public. As part of his research programme, he began to look at the new developments that were taking place in the poultry industry. With integrated links in feed provision, production and processing, not only was this concept applicable to broilers and battery hens, there was no reason why it could not be applied to the pig sector. Before long, Bill was lecturing to students on the basis that the size of an egg production unit ought to be geared to the carrying capacity of a 5 ton truck which could deliver food for ten thousand birds and collect the eggs from two units on the return journey. The media got hold of the story and it wasn't long before Bill started the first of his many local radio broadcasts based on this theme of so-called 'factory farming'. He revelled in the publicity. Production, however, was only part of the food chain. Just as important was marketing the product to the consumer; this was the common link throughout the meat sector. However efficient the producer might be, his profits depended on the costs of transport, slaughter, processing and marketing.

Following a study trip to America, financed by a commercial company, Weeks realised the enormous potential that was possible if only the industry could get its act together. On his return, and in conjunction with Mark Carpenter who was then Principal Agricultural Economist, he proposed to Mac that they should organise a national conference to highlight what was happening in the States. This was the platform Mac had been looking for, but he insisted that the conference must examine the whole of the meat industry. With his various contacts in the Ministry of Agriculture, Mac managed to attract sufficient funds to help launch the conference in April 1962, and to ensure that it attracted widespread publicity. Under the title 'Meat Marketing and the Farmer', the three day conference attracted over two hundred delegates. Bosanquet, in his Foreword to the published proceedings, stated that 'If this conference is to justify the hopes of its organisers this record must not be regarded as an account of something completed, but rather as a working document at the beginning of a long process of self-examination.' Mac Cooper gave the introductory paper and pulled no punches in his criticisms of the status quo.

We have not arranged this conference to make soup out of the bones of the £67 million skeleton now residing in the Minister of Agriculture's cupboard after last year's debacle in meat and fatstock trading. The decision to hold it was taken last May, long in advance of the events that have caused the Ministry to set up a committee of enquiry into meat. We were concerned with more general issues which we thought were bound to affect methods of distribution, as well as the nature of demand for the more conventional forms of meat — for example, the increase in home supplies of carcass meat, the mushroom growth of the British broiler industry, which now makes poultry meat more important than home produced mutton and lamb, the development of super-market trading, and, in a longer term context, Britain's possible entry into the Common Market.

He went on to underline that his purpose was to provoke the audience into thinking about the wider perspective of the problems which affected everyone in the industry, rather than their own self-interested sections. He had no wish to deny a reasonable level of profit both to the producer and to those who link the farm gate and the shop counter.

But what I do deprecate is the needless waste, the unnecessary double handling, the alternation of gluts and shortages, and the general inefficiency that characterises the processing and marketing of much of our meat. Unfortunately the deficiency payments as they are at present administered are not encouraging anyone to make the effort to repair the weaknesses. Truly they are deficiency payments in two senses, for they have also to cover the deficiencies in our marketing methods as well as the shortfall in farmers' returns, and it is the unfortunate taxpayer who is footing the bill.

Whilst Mac declared some dislike of the intensive methods of broiler production, he used this industry as an example of how a properly organised processing and marketing operation might benefit other sectors. Admittedly from a standing start and without all the traditions of the other meat sectors, broiler producers were nevertheless competing successfully for an increasing share of the housewife's meat budget. Furthermore, this was being achieved without any financial support from the taxpayer. He then returned to one of his favourite subjects, the livestock auction and the grading system on the hoof.

What is the point of grading on the hoof when there is an imperfect correlation between such an appraisal and ultimate carcase grade? Is there any necessity for the handling and double cartage, which do not add one-penny piece to the value of a beast, but cost a lot of money. If there have to be auctions to satisfy the gambling instincts of vendor and buyer, let them be on the cooling floor of a large and efficiently run abattoir which is located, not in the centre of some urban area, but in the heart of a meat producing area.

Mac spent some time advocating the removal of slaughterhouses from the centre of cities, Newcastle being a good example. Their relocation near to towns in the heart of the meat producing areas would not only ease urban traffic congestion, but would offer the opportunity to replace these inefficient and outdated establishments with modern meat factories, where animals could be dealt with humanely. This was increasingly the case in the pig industry, 'where such factories turned out pies, sausages and packaged meats which require the minimum of handling or cutting at the retail stage. It is an organisation which will serve the needs not only of the small retail butcher, but also of the large self-service type of store which require a standardised type of product.' He continued with some prophetic vision:

Whether we like it or not, there is bound to be a large increase in these super-markets, which will depend on centralised buying. The buyers for such organisations are not interested in assorted penny packets. They want to buy in large quantities, a standard article that is always true to grade. Today, apart from broilers, it is imported meat such as New Zealand lamb rather than home-produced which suits the needs of the multiple buyers. Until such time as home-produced meat is offered to him in quantity with a similar grading warranty as that carried by New Zealand lamb or Danish bacon, it will sell at a disadvantage as compared with imported meat. The implications of this point will become more serious as the importance of super-market trading grows.

Again with some foresight and thirty years before the onset of the BSE fiasco and its related health scare, he makes the telling point:

Not the least of the advantages which will come with large central slaughterhouses is adequate meat inspection and control of hygiene neither of which is possible at the present time because of the number and the dispersion of slaughterhouses. It is scarcely to Britain's credit that the American forces will not accept British meat because, justifiably, they are not satisfied with our slaughter hygiene. The same appalling situation is limiting the possibilities of our exporting fresh meat to the continent, especially to Germany.

The previous year Mac had been to the Argentine on an extended study and he told the audience about his visits to their export slaughterhouses. The scrupulous carcass inspection he witnessed was not only to ensure that the meat was fit for human consumption, but also (on Britain's insistence to safeguard our herds and flocks) that the animals were free from any symptoms of foot and mouth disease. He concluded: 'One cannot help wonder why similar precautions are not taken in our own slaughterhouses, to safeguard the health of human beings.' No one can say that the industry had not been warned.

The latter part of this wide-ranging paper concerned the need to re-examine the role of the Fatstock Marketing Corporation. It had been in existence for eight years and there were no obvious signs that this organisation was making real progress. He was wary of monopolies and certainly did not advocate a producer's meat board. He preferred to see a blend of farmer controlled co-operatives and privately financed concerns running the network of meat-processing plants, which he envisaged in the future. In his view: 'Each type of business will be good for each other in maintaining elements of healthy competition to safeguard efficiency and combat the real dangers of buyers' or sellers' rings.'

Finally he concluded that there was a strong case to introduce some form of central organisation, not unlike the Pig Industry Development Authority but with wider powers, to promote meat sales, investigate market trends and generally help the industry to develop.

The Milk Marketing Board has not only made a success of marketing; it has been equally successful in influencing husbandry and dairy farm management, thereby materially improving the efficiency of the milk industry. Meat production is in need of similar stimuli. In the modern context, marketing commences at the earliest point of production, not only to ensure the right sort of product, but also to ensure that it reaches the market in an orderly way with minimal fluctuations in supply. It is in this field as well as on the research side that an overall meat authority would be of greatest value.

Predictably, Mac Cooper's paper produced a heated discussion amongst the wide interests represented in the audience. The subject of whether there should be a Meat Board was uppermost in the minds of many of the delegates. No doubt some saw this as a universal cure of all their ills as they looked enviously across at dairy farmers. Most, however, agreed that the Milk Board model was unlikely to work with such a range of different products and species as were present in the meat industry.

There followed many other equally controversial papers presented by advocates of the new supermarket chains, traditional butchers and auctioneers in defence of the status quo, as well as practical farmers searching for guidance as to the way the meat industry should develop over the next decade. Even the poor housewife came in for a roasting during the discussion following Mark

Carpenter's contribution on a recent survey of changing trends in the demand for meat. A farmer delegate, Mr Donaldson, agreed that the supermarket expansion was unavoidable, but he was not convinced that meeting consumers' demand was necessarily a good thing.

> I have an absolute horror of the British housewife with a lead in anything. British cooking is the worst in the world. The British housewife spoils the vegetables in water; she says the broiler is tasteless because she does not know how to make sauce. The whole thing is really beneath contempt. I am horrified by the suggestion that we may be going to be led by the British housewife. She is already asking for the brown egg, which is twice as expensive to produce and is known to eat exactly the same as the white.

No doubt the comment was designed to lighten a discussion which was becoming rather academic, but as usual the housewife got the last laugh. Mrs Sykes replied: 'Speaking as a much maligned British housewife that my old friend Jack Donaldson has been attacking, there is a saying that nations get the governments that they deserve. Possibly husbands get the food that they deserve!'

The conference was an undoubted success and it stimulated a widespread public debate that continued for many months. It drew Mac into detailed discussions with the Ministry on how his ideas might be progressed and it led indirectly to his being asked to chair the newly formed Beef Recording Association the following year. The conference was also the spur to create a new Department under Carpenter. He became the first Professor of Agricultural Marketing in the country.

Another significant by-product of the conference was the contact with John Ashton, who at the time was employed by the Ministry in a fairly senior position. He had come to public notice as Chairman of the Royal Commission on the teaching of agricultural economics and had had some involvement with the planning and financing of the conference. He was an Oxford graduate and according to Bill Weeks, 'John had been an ex-Fleet Air-Arm observer on a Swordfish aircraft which helped sink the *Bismarck*. He was a downy bird and highly intelligent. One night he would be drinking champagne with his economist friends in Simpson's on the Strand, the next he would be holding forth in the Working Men's club at Ashington in Northumberland, about how the members might capitalise on their assets.' Bill also reckoned that there was much about Ashton that was a mystery, even to the point of his involvement with the British and American Intelligence network. Be that as it may, there is no doubting that here was a man who could bring real presence and credibility to Mac's team. Bosanquet also knew Ashton, so when the vacancy for the Head of the Economics Department became vacant, he was approached, persuaded and appointed without the position ever being advertised, a procedure, which would fall well short of employment and equal opportunity law today. But they got their man.

So with Professors Ashton and Carpenter respectively heading up revitalised Economics and Marketing Departments, here was the means by which student numbers could expand without constraint. They had plenty of space in the new building and with the help of the enigmatic Bill Weeks, they set about creating a name for themselves. Weeks remembers Mac presiding over these two new activities like a benign despot. 'He wanted to know about everything that was going on. He gave us a free hand, but God help us if we stepped out of line.' This was also an opportunity to interest the commercial world in the relatively new concept of food marketing. Led by Mac, the group quickly gained a reputation with the forward thinking companies, which led directly to funding for research projects and scholarships. This was new money and it provided much encouragement both to the Dean and to the Vice-Chancellor, as it confirmed that the changes in strategic direction which they had engineered for the Faculty had been well founded.

Over the course of the next few years, this diversification resulted in an annual increase in student intake from eighty to about two hundred. Inevitably it also meant that the original agricultural course and the farming focus within the Faculty became less influential. This was of real concern to Mac Cooper. He had always held that students should have a broadly based training with a generous portion of practical application thrown in. But he was also a realist and the move away from the general degree to a wider range of specialist degrees amounted to Hobson's Choice, if the faculty was to have a long-term future. As might be expected there was some disagreement between Cooper and Ashton as to the contents of the emerging courses in rural economy and environmental science. Mac was determined that they should not become too theoretical and he insisted that a significant element of farm management should be included in the syllabus so that students were discouraged from seeking refuge in ivory towers.

Graham Ross was one of the people Mac brought in to help ensure that this happened. A District Advisor with the Ministry of Agriculture, Graham had applied for a research scholarship at Newcastle. Although unsuccessful, Mac Cooper, as the chairman of the selection panel, spotted him as someone who had been working closely with farmers in the field and who also had a firm grasp of economic principles. Such men were few and far between. Accordingly Mac persuaded Ashton that Ross should be offered a post as lecturer in the Economics Department, a job for which he hadn't even applied. He accepted with some enthusiasm and stayed at Newcastle for the next seventeen years. Here was another example of Mac's influence through identifying people who had the strength of personality and the appropriate experience to ensure that the vision which he had for the new

157

Economics Department was not swept away in an avalanche of trendy modelling techniques and desktop theories.

In 1960 the Coopers moved to West Farm, Tritlington, a couple of miles down the road from Cockle Park. The University had bought this 200-acre adjoining farm and the land had been added to the main unit. The farmhouse had been completely renovated, making it into an impressive residence for the Dean and his family. By this time the girls were finishing their various nursing and medical qualifications, either in Edinburgh or in Newcastle, so it was only on relatively rare occasions that the whole family came together.

Mac and Hilary, however, were generous hosts, especially to impecunious postgraduate students who happened to be resident in the Cockle Park Pele Tower. Often a group of them would be invited for Sunday lunch or evening drinks. The Coopers also had many overseas visitors, especially from New Zealand, who as often as not would arrive unannounced and stay for weeks at a time. If they were young relatives they would often manage to find employment on the farm. Hilary was an expert in dealing with the unexpected. One evening an elderly couple arrived at the front door laden with heavy suitcases, obviously expecting to stay for some time. It took some skilful questioning to find out that he was a South African professor whom Mac had met on one of his world tours and had invited to visit any time he and his wife were in England. They

Mac loved talking to farmers and stimulating discussion.

158

were of course made enormously welcome; beds were aired, food produced and they stayed for a week.

This Cooper hospitality also extended indirectly to the students in Newcastle in that Mac gave permission for them to hold an annual barbecue at Cockle Park. This was a momentous occasion, with either a roast lamb or a pig the centre of attraction. It was an opportunity for academic staff, postgraduates, farm workers and students to participate in a real 'knees-up'.

Indeed, as a couple, Mac and Hilary were an integral part of the community at Cockle Park, attending harvest suppers, race nights and all manner of events. Hilary became thoroughly involved in the local community and even served a term on the local Council. She had a particular skill in identifying with the young wives of several of the research workers at Cockle Park, as she had done previously at Wye. Often they were a long way from their families and their single-minded men worked long hours, often including weekends at the farm. Hilary had had years of similar experience with Mac and knew how difficult it was for these wives to cope and to understand. But she did not confine herself to the postgraduate staff; she was equally welcome in the farm cottages, spending many hours in animated conversation with the wives of those who worked with the stock or on the land. There is no doubt that Mac and Hilary were a much-loved couple.

Although extremely busy with all his internal and external commitments, Mac also found time to take an interest in the College rugby side, being an active President and consultant coach. He would regularly attend the annual dinner and notwithstanding the reverence in which the gathering held him, he could be counted upon, provided he had consumed sufficient whisky, to recite at least some of the verses of 'Eskimo Nell'. Next morning it was business as usual. There was never any question that the previous night's festivities diluted the priorities of the day.

His lasting legacy to Newcastle on the sporting side was the acquisition of Close House, a 70-acre private estate, about five miles to the west of the city. It was rumoured that he authorised the purchase without the sanction of the University authorities and almost got fired as a result. In the event, it was probably one of the best deals ever negotiated and now provides the students with a comprehensive complex of sports pitches and even a golf course.

During the sixties the Cooper family began to extend as the girls married and started their own families. Diana was the first to leave the nest, marrying Martin Thompson, an arable farmer from Lincolnshire. This was an occasion for an international reunion, with Hilary's sisters from Ireland and South Africa as well as Mac's two sisters from New Zealand, coming over for the wedding. Barbara, now a qualified doctor, married John Craven at the end of 1962. He had just completed his PhD at Cockle Park and they set off for London, as he was to join the Milk Marketing Board as a trainee farm management consultant.

Dai Morris, an ardent Welshman, graduated from Aberystwyth in the early sixties. Having heard about the enthusiasm being generated at Newcastle, he decided to come north to do his postgraduate work, having first obtained a Ministry scholarship. In fact he happened to share a room in the Pele Tower with John Craven as well as many hours in the Oak Inn. He was also responsible for a considerable amount of ribald comment concerning John's friendship with the Prof's eldest daughter. There was even a suggestion that this was the only way that he was going to get his higher degree. Fate, however, provided an opportunity for revenge. After Dai had finished his PhD, Mac asked him to stay on at Cockle Park as Assistant Farm Director. In due course he met Cynthia (or Squint as she is known to the family), the remaining unmarried daughter. The rest, as they say, is history. It can't have been easy continuing to work for his father in law, once Dai had also joined the family, but there exists correspondence that proves Mac had a very high regard for the work which he did in managing the farm. But Cockle Park was a stepping stone to bigger things and within a few years the Morrises had moved to Wiltshire where Dai was to manage the two and a half thousand acre Bowood Estate.

Farming in the sixties might be described, in retrospect, as a golden age. Following the 1957 Agriculture Act, where the principle of standard quantities were introduced to control subsidies, it became clear that efficient farmers could continue to make money at price levels which were guaranteed ahead for three years. This was also the period during which the technological progress in farm machinery and specialised building design developed a few years earlier, enabled progressive farmers to modernise their systems and take advantage of the generous grants, as well as tax allowances, available on capital investment. It was a time of application, whereas the previous decade had been one of innovation. Even the politicians urged agricultural expansion with the slogan 'Food from our own Resources', aimed at encouraging home production and replacing expensive imports. This period also coincided with the arrival of the farm management adviser, or consultant as they chose to call themselves in later years. The Ministry of Agriculture continued to provide extension workers or advisory staff without charging fees, as did the rapidly growing commercial firms who supplied their farmer customers with feeds, fertilisers and chemicals. But other organisations, notably the Milk Marketing Board in the public sector, as well as an increasing number of private individuals, launched fee paying services in response to the availability of farm recording grants introduced to encourage farmers to apply business principles to their enterprises. It was the start of the concept of gross margin analysis, whereby simple enterprise costing helped to reveal the loss leaders and whether it would be worth expanding one operation at the expense of another.

It has to be said that whilst this development heralded a major step forward in applying financial evaluation as an aid to decision making, it also led directly to disaster for those who did not understand the need to assess overhead costs, balance sheets and cash flows. This was especially true if substantial borrowing was part of an expansion plan. Farm management or extension work therefore became a respectable occupation and the forum for endless debates at conferences and farm study groups up and down the country. Unlike any other industry, farmers were keen to share their experience with others. They welcomed research workers and visiting groups to their farms to demonstrate new techniques and developments. They were happy to show their financial figures to each other and with so many advisers involved, there is no doubt that this period saw a significant improvement in the exchange and implementation of technical knowledge.

This momentum led to the increasingly intensive farming practice of the latter part of the century. The emphasis was on higher output, either per animal or per acre through more generous feeding or by using heavier dressings of artificial fertiliser. New varieties of crops were introduced; improved selection in livestock breeding programmes was initiated and foreign breeds of beef and dairy cattle began to be imported. Production systems designed to fit the farm resources were also identified and demonstrated to farmers. Stocking intensity became a watchword so that more cows, sheep and beef were carried on each acre. Winter housing, especially for dairy cows, moved rapidly from cowsheds to cubicles, self-feed silage and milking parlours. Pig and poultry units occupied buildings, which controlled their environment, and even sheep were housed on some heavier land farms to increase the carrying capacity of the spring pasture.

Similarly on the arable side, fertiliser use coupled with selective weed-killers and insecticides pushed up yields per acre. Added to this, more powerful tractors and specialist cultivating and harvesting equipment enabled crops to be managed at optimum times, rather than when the weather dictated. The acreage of cereals expanded rapidly at the expense of poor permanent grassland. The 'Barley Baron' entered the vocabulary. In other words farming was ceasing to be just a way of life; it was becoming a highly capitalised and technical business with the gap between the leaders and the laggards becoming ever wider. The results were dramatic by any standards. Food production increased, imports fell and small farmers left the industry in droves. As machinery replaced labour, farm workers lost their jobs and land values started a climb that has, with a few fluctuations, continued ever since.

On the downside, the new problems of pollution through slurry and silage effluent; the ripping out of hedgerows to enlarge fields so that bigger machines could operate more effectively; and the many animal welfare issues associated with intensive livestock production, had all yet to become major public concerns. In addition, it was perhaps inevitable that hard on the heels of the euphoria of expansion and investment, there would be a sharp reminder that whatever goes up must come down. This was to arrive in the form of the oil crisis and

161

galloping inflation in the late sixties and early seventies, followed by entry into the European Common Market.

So whilst throughout Mac Cooper's career, the fortunes of the agricultural industry had continued to move up and down like the proverbial yo-yo, it is probable that the sixties saw the most spectacular change. This was exciting because it provided a climate within which farmers were keen to embrace opportunity. They were looking for guidance and direction. It was a rewarding time to be working in the industry, in the Faculty or outside in the field.

As well as the diversification into the new marketing and economic spheres, a number of other personalities began to make their contribution at Newcastle as the sixties gathered pace. Prominent amongst these was David Armstrong. David had completed a PhD at Newcastle and then went on to the Hannah Dairy Research Institute in Scotland to work on ruminant nutrition under the much-respected Kenneth Blaxter. Armstrong returned to a personal chair at Newcastle in 1963 and headed up the Agricultural Biochemistry Department. He subsequently developed a highly successful nutrition unit in a converted farm building at Cockle Park, where he researched techniques in the study of quantitative digestion and metabolism in ruminants using cows with re-entrant canulae. This was pioneer work and quickly established Armstrong as one of the leading world authorities in his field. Here was yet another example of someone recognised by Mac Cooper as a scientist with a major contribution to make and he made sure that the resources he required were made available. Just as importantly he ensured that he had the freedom to develop his work with the minimum of interference.

It was also at about this time that another energetic enthusiast appeared on the scene. Looking for a job after he had completed his PhD at Nottingham, John Prescott attended an interview at Wye with Principal Skilbeck one day, followed by a walk round Cockle Park the next with Mac Cooper. He had met neither of them before, but his account of the interviews confirm the impression previously formed, that these two characters could not have been more different. The first meeting with Skilbeck was very formal and rather intimidating, John being told that there was a job available, but it was for a limited period of two years only. He would be lecturing with little prospect of becoming involved in research work. Take it or leave it. Mac, on the other hand, sat with his legs dangling over a laboratory bench and chatted amiably over a cup of tea. The vacancy at Newcastle had arisen as the dairy lecturer had gone overseas on a sabbatical year, and Mac wasn't sure that he would come back, so there was a good chance of a more permanent position, provided things went well. After a walk round Cockle Park, John was left in no doubt that Newcastle was the place to be.

In fact, the wheel came full circle. Prescott, after working a number of years with David Armstrong on the dairy side and Mark Carpenter on meat marketing, left for a two-year sabbatical in Argentina. After he returned to Newcastle in 1973, Mac Cooper had moved on and John felt the time was right to move as well. He went to Edinburgh and after only four years took the Chair of Agriculture following Frank Elsley's untimely death. In the early eighties he was headhunted for the job of Director, at what was then the Grassland Research Institute at Hurley. The full circle came with his last appointment as Principal of Wye College in 1988 on the retirement of Ian Lucas. It was ironic that John should start his career turning down an offer of a job at Wye, come under Mac Cooper's influence and finally end up there as Principal, no doubt much to Mac's satisfaction.

Wye College has recently merged with the University of London Imperial College of Science, Technology and Medicine. The forward vision is that agriculture, food, and the environment, with their relationship to human health, will be the focus of the future. This merger will create an organisation that will have both the resources and the international credibility to become a world leader in the field. Even now Wye is forging ahead with its 'Distance Learning' programme. This initiative targets the thirty- to forty year-old mid-career professionals who want to gain further qualification without recourse to full time study. Courses in Business Management, Bio-diversity, Environmental Assessment and Sustainable Agriculture are currently provided to a thousand students in over one hundred countries. The costs in real terms of achieving a postgraduate degree at Wye by these means amounts to about one third of that compared to the conventional approach in residence. Furthermore, students can continue with their employment and add value to their work experience. It seems that Wye can indeed look forward to an exciting future.

The mid-sixties were a time of wider recognition for Mac Cooper. He no longer had to prove himself. The politicians accepted many of his opinions and the demand for him to serve on public bodies began to increase significantly. It was in his nature to accept the majority of invitations that came his way and with hindsight, he ought to have been more selective. Such tasks were always cumulative; everything else had to be done in addition. As well as joining the Minister of Agriculture's Scientific Advisory Panel (the incumbent Labour Minister being Fred Peart), Mac was appointed as the chairman of the newly formed Beef Recording Association. This was a Government and industry supported initiative arising indirectly from the marketing conference held previously at Newcastle. Ken Baker became the Association's Chief Executive and recalls Mac's persuasive methods to recruit him. Ken had been an undergraduate at Newcastle in Wheldon's time. He had applied for a postgraduate scholarship under Professor Martin Jones' supervision at Newcastle and although most of his work had been undertaken at the Grassland Research Institute at Hurley, Ken, a qualified pilot, often borrowed an RAF plane to fly up to Newcastle to discuss his work.

Ken had first met Mac in the late fifties, when as Secretary of the British Grassland Association, they worked together in Mac's presidential year. Shortly after it became public that the new Beef Recording Association was to be set up, Mac paid a visit to Hurley, followed by an overnight stay with the Bakers. After the consumption of a large quantity of whisky, Ken woke up the next morning to find that he had agreed to become the Association's first Chief Executive. It sounded a rather impressive title, but in fact Ken was the sole employee when he took up the job in March 1964.

The aims of the Association were to establish performance targets for the various beef production systems and to help set up recording on both commercial and pedigree units. In addition there was a need to help establish properly managed central test stations so that bulls were selected for use in AI on the basis of their own live weight performance or through conventional progeny testing.

It was a difficult task as the members on the Committee included uneasy bedfellows in the form of representatives from the Ministry, commercial AI organisations, the National Cattle Breeders Association and the National Farmers' Union. According to Ken Baker, Mac Cooper was the ideal chairman. He let everyone have his say and then steered the decision-making expertly in the way both he and Ken had previously agreed. Amongst these vested interests, Mac was seen as truly independent, and above all he managed to attract the necessary funding in order to keep the thing afloat.

Following on from the protracted debate about a national Meat Authority, the 1967 Agriculture Act finally established the Meat and Livestock Commission with a remit to co-ordinate activities in the whole of the meat sector. This meant that there would be logic in merging the Pig Industry Development Authority and the newly formed Beef Recording Association, with the intention of including sheep in due course. Mac, however, was anxious to ensure that the pig sector, which already had an infrastructure of field staff distributed round the country, did not simply absorb the enthusiasm which he had initiated with the various players in the beef world. Indeed some of his own committee preferred to keep the Beef Recording Association as a separate independent entity, rather than agree to let it be funded by a levy from the industry. So it was with considerable skill that he kept his colleagues on board and as a result, the activities came together in a constructive way, with Ken Baker being appointed as Livestock Director of the MLC. Once Mac had achieved this objective he was content to ease out of the picture and let others get on with running it. But in Ken's view, the Beef Recording Association and its successful integration within the MLC would never have got off the ground had it not been for Mac Cooper's determination and vision, coupled with his ability to knock heads together, if the situation demanded it.

It was with much satisfaction to all his friends and colleagues, that in the birthday honours list the following year, Mac Cooper was awarded a CBE for his services to agriculture. Here was public recognition that his crusade against outdated practices and tunnel vision within the industry was acknowledged. And more important, his efforts directed towards proposing practical improvements were appreciated. Needless to say the honour was duly celebrated in style, both in Northumberland and in New Zealand.

Further recognition came in 1970 when Mac Cooper added the coveted Massey-Ferguson award to his list of achievements. The presentation, made by Jim Prior, the Conservative Minister of Agriculture at that time, highlighted Mac's contribution in terms of helping farmers towards a better understanding between grassland and grazing animals. The citation went on: 'Through his outstanding capacity to inspire others he has been pre-eminent in the vital task of presenting the findings of agricultural research to farmers in terms of economic advantage.'

Back in Newcastle, however, things were changing. The balance of student numbers began to swing more away from the Agricultural Department towards Marketing and Rural Economy, so the task of keeping abreast of all that was going on in the Faculty became more difficult. With all his outside commitments,

Mac was happy to talk to local Grassland Societies, especially if the perk was a ticket for Cardiff Arms Park.

Mac was often unavailable for weeks on end. Where there had been murmuring in previous years, it had been from the old guard, seeking to turn the clock back. Now it was the new breed, headed at Professorial level by ambitious and able people such as Ashton, Carpenter and Watkin-Williams, a plant geneticist. These were men who were stirring the pot and the discontent was becoming apparent to Mac and it caused him some concern.

In 1968 Bosanquet retired as Vice-Chancellor and was replaced by a medical man, Henry Miller. Agriculture had, under Bosanquet, enjoyed a long period of preferential treatment and Mac had had more than his fair share of the available financial resources. Priorities were about to alter, and although still a senior member of Senate, Mac no longer had the sympathetic ear of the Vice-Chancellor. One example of the change was in the composition of the Farms Committee, which reported back to the University Council on their progress and accounts. Whereas under Bosanquet, the composition of this committee had been agreed with Mac and included local and prominent friendly farmers, Miller insisted that the appointments were reviewed and the members invited to serve by the Council. As a consequence, they were drawn from a wider cross-section and were not prepared to accept the assurance that all was well without solid evidence. In fact all was not well on the farms, especially at Cockle Park, where substantial losses began to accumulate as the costs of the ongoing research work had escalated. Results at Nafferton were not much better and there was talk at one time of giving up the tenancy to realise capital and save overhead costs. Rumour had it that a number of interested parties close to the Farm Committee members were keen to see the farm come onto the market. But common sense prevailed. In the event the Council had little option but to write off these losses, but the outcome was a much more rigorous approach in the way research work was accounted for and justified in the future. Gone were the days when a member of staff came up with a good idea and needed twenty ewes to carry out a trial. Now it depended on who was going to find the money to pay for it. Efforts were needed to attract externally commissioned projects from commercial companies and staff attitude began to lean more towards work that had a chance of sponsorship. Mac didn't like this trend as it interfered with his own judgement as to what was important both to research and to its application on farms. What might seem important to companies who had the funds to invest and an interest in selling commercial products, might not be in the best interests of their farmer customers.

The approaching downturn in agricultural fortunes did little to help the financial results on the farms. In the early seventies the oil producing countries cut production in an effort to force an increase in world prices. This had an immediate effect on the price of fertilisers, feeding stuffs and machinery, without any adjustment to farm product prices. The impact in cost escalation spread like wildfire throughout the whole economy and inflation took off to levels in excess of twenty per cent. Not surprisingly, there were serious consequences for many farmers who had borrowed heavily in answer to the Government's call for expansion. This crisis was followed quickly by Britain's entry into the European Economic Community with its attendant Common Agricultural Policy. This provided farmers with a very different support mechanism based on buying up surpluses when the market became over supplied. These intervention stocks were then stored at enormous cost and fed back into the market if the surplus reduced, or sold at giveaway prices on the world market. The astronomic cost of this policy was partly financed by charging a levy on food imports from countries outside the EEC, although as production continued to increase as farmers tried to cover overhead costs, a higher proportion of the burden fell on the taxpayer.

The picture was further complicated, and still is, by the different and fluctuating currencies in each of the member states. In an attempt to treat farmers fairly throughout the EC, a system of monetary compensation was introduced, enabling farmers within each country to be recompensed should exchange rates deviate significantly from their European neighbours. Triggering this mechanism, however, depends on national Government initiative, which is inevitably viewed in relation to the rest of that country's economy. Only when and if monetary union becomes a reality, will British farmers feel they are competing with their European counterparts on equal terms. But arguably of more significance than all of this was the fact that when Britain joined the Common Market, decisions with regard to agricultural policy and financial support passed inexorably from Whitehall to Brussels. It was now up to the UK Government to argue the case for its farmers. Farmers had lost their direct negotiating influence with Government.

The prospect of Britain joining the Common Market was viewed with serious concern in New Zealand as exports of butter and lamb to the UK constituted a significant part of her overseas trade. On a visit to Australia to attend a Commonwealth University congress as Newcastle's delegate, Mac Cooper took the opportunity to visit New Zealand both to look at their current research programme and to visit his family. He also gave a number of interviews and talks with regard to the forthcoming move towards Europe. On his return he wrote a local newspaper article about what he had seen during his trip.

I had expected, when I returned to my native New Zealand earlier this year, that I would find a depressed farming community. Doubtless my thinking was conditioned by the current pessimism in British farming with high interest rates and accelerating costs of production which have not been matched by increases in prices of the principal farm products. I had also anticipated that New Zealand would be feeling the effects of the curtailment of its traditional markets for dairy produce and meat, as a consequence of the United Kingdom's membership of the EEC. I could not have been more mistaken. There was occasional gloom – farmers would not be farmers even in paradise if they did not have a grumble – but the general impression was one of optimism which went well beyond the very characteristic 'she'll be right' attitude even in the face of adversity.

Mac went on to outline the way in which the New Zealanders had been anticipating the need to look elsewhere for new markets and they had made excellent progress in such places as Japan, South Korea and America. In addition, some were in the process of diversifying their farming activities on suitable land away from mainstream enterprises such as sheep, beef and dairy into fruit growing, viticulture and deer farming. In other words here was an example of an industry which was not dependent on the State for financial support and as a result did not waste time complaining that they needed more help. They simply got on with the reality of the marketplace. This was a cause Mac had long espoused and his conclusion, sadly, was that Britain would find it even more difficult to move towards an unsubsidised agriculture, even if she wanted to. The UK, farmers were too small an electoral entity to impress the politicians, whilst this was manifestly not the case in countries such as France and Germany, with their hundreds of thousands of small farms. It looked as if subsidised agriculture, with all its warts, was here to stay.

Malcolm Willis, who had been a student when Mac had arrived in Newcastle, returned as a lecturer in the early seventies to take over from John Prescott. After his postgraduate degree in Animal Breeding at Edinburgh, Willis had joined the Milk Marketing Board for a six-year period before applying for a job in Cuba to work with a man called Reg Preston. Preston, who himself had been a PhD student at Newcastle and was about to join the staff when Mac Cooper arrived in 1954, had, according to Willis, been told by Mac to 'buzz off and get some experience somewhere else'. In his view Newcastle was far too inbred already. Preston reckoned that was the best advice he ever received and he went on to make his name at the Rowett Research Institute in Aberdeen with his work on the early weaning of calves and later with the development of the 'barley beef' system. He then progressed to Castro's Cuba, where, although he found the communist bureaucracy frustrating, the teaching and research facilities were very good. Willis was delighted to join his teaching staff.

During his time in Cuba, Preston had invited Mac Cooper to act as external examiner to their postgraduate students and as might be expected he had little difficulty in persuading him to make the trip. In fact he enjoyed the experience so much he repeated the visit on a later occasion. Thus it was that Willis got to know Mac quite well and when he decided to return to the UK, felt it worthwhile to write and see if there were any vacancies at Newcastle.

Willis' request happened fortunately to coincide with Prescott's imminent departure for his sabbatical in Argentina and he promptly stepped into a two-year temporary appointment, without even having had a formal interview. Willis found the Faculty very different to the one he had known as a student some fifteen years previously. It was very much bigger with the Agricultural Department having been outgrown by the seeming multitude of peripheral courses. There were many new faces, a new building and Newcastle was expanding as a University in its own right. Perhaps the most striking impact, however, was the fact that Mac Cooper seemed a rather remote figure. He had by now been appointed as a Pro-Vice-Chancellor, which meant his close involvement with University affairs was less directly related to those within the Faculty. It was also obvious that the management of the farms and their financial viability was under constant pressure. Furthermore, there was a noticeable undercurrent of discontent amongst some of the senior people who thought there ought to be some fundamental changes at the top. It was all a bit of a shock, as Willis had found Cooper to be on top form when he had recently been in Cuba. But it was even more of a shock, especially to his new colleagues, when it became known in 1971, that Mac Cooper had resigned, and intended, at the age of sixty-one, to take up an appointment with the World Bank and move to Spain. Here was the chance some of his antagonists had been waiting for, principally the opportunity to abolish the position of permanent Dean and allow the leadership of the Faculty to rotate. This, they argued, would enable the other, now bigger, Departments to have more influence in the future.

After Cooper's departure the following year there followed much discussion at University level as to a new structure. The decision was indeed that there should not be a permanent Dean, but that the position would be held by a senior individual within the Faculty who would be duly elected by his colleagues and serve for a period of three years. As far as general Agriculture was concerned, a new Professor would be appointed to head up the Department and take responsibility for the farms. This latter position was seen to be of critical importance, especially by those who had a responsibility for the financial performance of the farms, and it became clear that their preference was to lobby for someone who had an impressive track record in this field.

The job eventually went to Gordon Dickson, a former Newcastle postgraduate who had only recently taken over as Principal of the Royal Agricultural College at Cirencester. Dickson had made a name for himself as manager for the Duke of Norfolk on his 3,000-acre estate at Arundel and despite having to upset his new employers at Cirencester, he opted for the challenge of Newcastle with enthusiasm. The appointment did not please everyone as the more scientifically minded academics felt that important as the farms were, other means could have been found to solve their financial problems. What was needed, in their view, was a professorial replacement for Mac Cooper who could continue in public where he had left off and at the same time re-invigorate the research work at Cockle Park to attract postgraduate students of high calibre. To be fair to Dickson, the first demand was simply unrealistic.

Mac had established a reputation throughout the country as one of its leading agricultural gurus. Whoever replaced him would have had an impossible task in trying to follow in his footsteps. With regard to their second priority, some felt that Dickson's leadership was not going to be scientifically inspirational and as so often happens in these circumstances, voted with their feet.

In the event the farms did begin to improve but it was a slow process. Dickson ensured that their activities were properly accounted for against the work they undertook on behalf of all the other departments in the Faculty. He also managed to get hold of two very capable young managers who were able to implement change and he obtained the funds necessary to modernise the outdated buildings.

The revolving Dean saga gradually became something of a political hot potato. Elections were soon replaced by 'appointments after consultation' and in later years as the job became more administrational, it became a burden which many of the more able staff were anxious to avoid. This seems an unfortunate outcome to the obvious frustrations felt at the time of Cooper's departure. Whilst to find someone who could provide an ongoing leadership and at the same time unite the Faculty might have been very difficult, nevertheless in retrospect, it rather looked as if the baby had been thrown out with the bath water.

Most commentators would agree that the most influential period of Mac Cooper's career came to an end after nearly eighteen years at Newcastle. He had made a huge impact in the north-east and was even referred to as an 'adopted Geordie' by those locals who themselves felt this to be the supreme accolade. During the first part of his tenure he projected Newcastle as the most exciting Agricultural Faculty in the country. He attracted ambitious and talented postgraduates, many of whom went on to fill senior positions in the industry both in the UK and overseas. Without doubt his influence with students, staff and outside within the farming community was remarkable and many still refer to his having had a profound effect on their future careers and indeed their lives.

10

SPAIN

The news that Mac had decided to leave Newcastle broke in the autumn of 1971. It came as a complete surprise to everyone, not least the University authorities that only the previous year had appointed him as one of two Pro-Vice-Chancellors. They were not best pleased that they had to repeat the rather arduous job of finding his replacement. There were other organisations similarly affected, notably the British Association of Animal Production (now Science) who had recently appointed Mac as their President. His departure was to mean that he could only play a nominal role in the Society's affairs with consequent embarrassment to his sponsors.

In the middle of a letter home on 29 September, the twins' thirty-first birthday, he wrote to Thelma, Joe and the family, prior to the news becoming public.

And now for the really big news. In six months I shall be leaving my present post to become Chief Research Co-ordinator in a World Bank development project in Spain. It is the first time that the Bank has lent money for such a purpose and naturally it is concerned that the money is wisely spent. The proposition came from out of the blue and the weekend before last we had a visit from a very nice American, Sam Freiberg, a Vice President of the Rockerfeller Research Institute. They are acting as agents, and Sam was casing me in comparison with two Americans who were also being considered for the post. Evidently he was satisfied and he went on to Spain where there was a unanimous opinion that I should be offered the appointment. This came by phone from New York on Thursday last and since then we feel we are in dreamland. The girls thoroughly approve though naturally they are sad that this part of our life will be coming to an end. We will keep our home going here, but Hilary will be spending most of the time with me in Spain, except for the hottest months of the year. The salary is just about double my present

171

one and it is income tax free. The contract is for five years and it takes me a year past my normal retiring age. There is a free house and a chauffeur-driven car (for work only). But it is not the material aspects alone. I am looking forward to one final big challenge in my life and this is the one that is going to be really satisfying.

He omitted to mention in his letter that payment would be in American dollars (a distinct advantage in those days) and that his pension entitlement would improve threefold. It is clear from this letter that both Hilary and Mac were excited about the prospect of a new life in Spain. Certainly the money at this stage in their lives was tempting, especially as Mac had not really thought much about his financial position as retirement began to appear on the horizon. They didn't even own a house, although by now they had moved fifteen miles further north from Tritlington to a rented property in Lesbury, owned by the Duke of Northumberland. This quite large stone house, on the outskirts of the village, stood in a beautiful setting overlooking the river Aln and within a couple of miles of the spectacular Northumbrian coastline. Holme House was an ideal location for family gatherings at Christmas and summer holidays for their nine young grandchildren, not to mention their daughters and sons-in-law. They had every intention of retaining the tenancy during the time they were to be in Spain.

Apart from the money and the challenge, Mac was also ready for a change. He knew there was increasing unrest about his position as permanent Dean.

Holme House, Lesbury, Northumberland.

He also knew that it wasn't possible, unless he fundamentally altered his priorities, to acquire the detailed knowledge to keep abreast of all that was happening throughout the Faculty, as well as keep tight hold on the farm operations. To have any hope of getting it all back on track, he would have had to sacrifice much of his outside work and such a prospect would have been unthinkable. He had given the best years of his career to Newcastle and maybe it was time to make way for a younger man. This new job was a golden opportunity, in more than the obvious sense, to get away from it all. Finally, Hilary was also probably keen for a change. The petty politics of the University was an irritant that she could see was undermining Mac's authority and reputation. His health was not good and the prospect of his avoiding the winter in the north-east must have been an added consideration in favour of the move.

Interestingly, the initial mention of Mac Cooper's name in the context of the World Bank job came from his old friend McMeekan. Following his term as Director of Ruakura Research Station, and his failed bid to become Director-General of Agriculture in New Zealand's Ministry of Agriculture, McMeekan had joined the World Bank as an Associate Director in 1962. He already had a formidable reputation in advising Governments with regard to their research and extension priorities. His work for the Bank involved the evaluation of development programmes, mostly in third world countries, and the active promotion of lending on agricultural projects. After spending a period in America, he returned to his farm in New Zealand, but continued to act as a

Family holidays at Holme House.

part-time consultant on overseas missions. One of these assignments was the project in Spain. Gordon McLauchlan, in his biography of McMeekan, describes the background to it.

> In the late sixties, the Spanish Government asked the World Bank to help in setting up organised agricultural research. McMeekan moved in with a high-powered team and decided to take a commodity approach. The main Spanish commodities – citrus, soya beans, and rice – were to be worked on individually within the context of the Spanish physical environment by a team of World Bank experts; and Spanish students in the appropriate disciplines were to be sent overseas for three or four years to get top-class degrees.

The theory seemed fine, but in practice the project was fraught with difficulty. Whilst at Government level, the loans and the conditions were approved, when it got down to implementation in the field, there were endless problems of procrastination and downright dishonesty. Some of the Spaniards just could not accept that anyone from outside could come and tell them how to improve upon the methods they had been practising for years.

How much of this McMeekan knew when he first raised the possibility of Mac becoming involved as Chief Co-ordinator is not known. However, McMeekan was a persuasive individual and clearly needed someone he could rely upon to make things happen. One of the issues, that might have given an inkling of what was to come, was the insistence that the Chief Co-ordinator for the project must have a doctorate. Mac had virtually every other qualification it was possible to have, but the DPhil to which he had originally aspired at Oxford had failed to materialise. This was a hurdle which McMeekan quickly overcame by his proposing to Massey (their former University in New Zealand) that the award of Honorary Doctor of Science might be an appropriate recognition for one of its most illustrious sons. This may well have been in the pipeline in any event, but according to one source, the arrangements all seemed to be rather hurried. Meanwhile the Royal Agricultural Society of New Zealand had invited Mac to undertake a lecture tour in May 1972, shortly after he was due to start work in Spain. During this three-week trip, it was agreed that he would be presented with his doctorate at Massey. A letter to sister Thelma is full of excitement about his forthcoming trip home, although he expected it to be a very rushed visit, as he had to be back in Spain by early June. He explained: 'There are a lot of critical decisions to be taken and I don't want too much to be decided in my absence.'

> I start the Spanish job about 15 April. I am looking forward to it and my appetite has been whetted by a brief visit there last weekend. There will be plenty of headaches, transacting business largely in another language,

with people with a very different background from the one to which I have become accustomed. But I am not daunted by the prospect of it all. The hardest part will be leaving the University and Cockle Park because these have meant so much to me the last seventeen years. It would have been so much easier if my present job had not been such a satisfying one.

It seems that McMeekan was about to return to New Zealand after a period of residence in Madrid and leave Mac to get on with the job. His comments also suggest, even before he started, that Mac foresaw some difficulty with both the language and the attitude of the people he was going to have to work with.

I am taking over the lease of the apartment that the McMeekans now have until such time as we have found our feet there and decide how we want to live. I can't think that Hilary and I can reconcile ourselves to an indefinite spell of flat life with no access to a garden. Mac [McMeekan] is keeping my seat warm meantime and this is good because he is a

Award of Honorary Doctorate – Massey, 1972.
Sir Alan Stewart, Vice-Chancellor, on the left of the picture.

hardheaded customer and he will soften up the Spaniards to the point where they will feel that I will be a welcome relief.

Although Hilary had decided not to accompany Mac on his New Zealand trip, this changed at the last minute owing to a cancelled operation that she had been planning to have. In April she wrote to Thelma:

As you can imagine the past few weeks have been traumatic, especially for Mac with all the farewells, finishing his job and hardest of all leaving Cockle Park which is looking very lovely with the spring grass beginning to come away. Having made the decision I feel a different person and tremendously excited. I'd been dreading this summer parted from Mac. So you'd better start stocking up with tree tomatoes.

In fact they didn't travel together as Mac had to go via New York for a briefing meeting, but he was delighted Hilary had decided to join him. There were lots of family to meet and old friends to visit, with the highlight of her being able to join in the festivities when Mac received his Honorary Doctorate at Massey. He described this, according to one newspaper article, as 'one of the most delightful things that has ever happened to me, and the greatest honour I have ever received.' Hilary was quoted as saying that when Mac had first heard that he was to be given this award from his old university 'he did not stop smiling for a week'. Irrespective of the Spanish dimension, it clearly meant a great deal to them both.

Thelma and Joe received a letter from Mac after they had been in Spain for a couple of months and it contained the first signs of the trouble that was brewing. He wrote from their apartment in Fernandez de los Rios in Madrid.

It is strange living in a seventh floor flat with the constant noise of traffic coming up from below, but it is quite remarkable how quickly we have settled. Hilary finds Madrid exciting and she has the added interest of learning Spanish. She is enrolled on an intensive course where English is never spoken and though it was very tough for a start she is now making a breakthrough. Unfortunately I can't find the time to take regular lessons but I must do so because a lack of understanding of the language puts me at something of a disadvantage in my job.

This is not going as well as I had hoped. Nobody seems to be able to make a decision and the amount of farting around over trivialities has to be seen to be believed. Sometimes I think I am living in a land of make believe. The redeeming feature is that the Spanish are such nice people and the wine is good and cheap.

The rest of the letter described their first impressions of the country, but did not record what must have been a considerable shock to both of them. McMeekan, who had set up the whole project and who was to continue to help keep it on track, had recently drowned in a boating accident back in New Zealand. Apart from the loss of a valued colleague, this came as a severe blow to Mac Cooper, for they had been good friends ever since they had met at Massey some forty years previously. He wrote in an international magazine: 'The grief for his passing which is felt in Spain and elsewhere is not limited to his scientific colleagues or men in high places who have known and respected him. It is shared by secretaries, porters, cleaners, farmhands and all manner of people who have known him and loved him for his great humanity and friendliness.'

To understand the problems which Mac and his team faced when trying to implement the project with the senior research staff in Spain, it is necessary to describe some of the background to their own appointments. John Gerring and Neil Worker, both New Zealanders, were involved in a supervisory capacity on behalf of the World Bank and visited the country on several occasions. John's job was loosely described as an 'itinerant troubleshooter', being called in by the Bank when things weren't going according to plan. They recall that some of the Research Centres had long traditions and were not organised in a way that Cooper and his staff were used to. They were part of the cultural heritage. The senior posts were passed down within local aristocratic families and the Directors ran these stations like fiefdoms. Nobody interfered with

McMeekan (standing) negotiating contracts on behalf of the World Bank.

them. They were completely autonomous with a complex infrastructure of privileges and perks for those involved. Some even had close relatives in senior positions within the Ministry of Agriculture. Furthermore, there were frequent changes within INIA (Institute of National Investigation Agrarios). This meant that not only did the President change, but all his subordinate staff moved around as well. Each time this happened, the new team took months to get to know the people involved in the project. In addition, they demanded a reassessment as well as a new budget, all of which had to be formally published in a report to be submitted to their headquarters in Madrid. It took endless hours of unproductive work, especially as the research priorities were unlikely to alter in the short term.

McMeekan had vast experience of this sort of bureaucracy, both in Spain and in Latin America, and one of his aims had been to persuade the respective Government officials that this must change. Their programmes could only succeed if there was a root and branch re-structure of their research organisations, backed up with continuity of leadership at the top. The Bank loans depended on this being implemented. The civil servants hung on his every word and everyone was in full agreement with what had to be done. Unfortunately the responsibility for communicating these decisions to the people who were supposed to make the changes, seemed not to rest with anyone.

The overall project focused on the major food commodities and each research centre had a resident co-ordinator employed by the World Bank who was responsible to Mac Cooper for ensuring that the agreed work was carried out in conjunction with the Spanish Director of the station. When Mac began to

Mac with a Spanish colleague, Madrid, circa 1973.

178

visit the different centres and meet his own team, he realised that some of these Directors had no detailed plans with regard to the project that had been agreed in Madrid. Furthermore they appeared to have little intention of changing their ways. Predictably Mac became increasingly irritated and frustrated, and the position worsened immeasurably when McMeekan's influence disappeared. Cooper tried to be polite at first, but he was a senior figure and he wasn't used to being back-chatted. He had obvious difficulty with interpretation, but no one could have been in any doubt about his body language.

Peter Mortimor, a Yorkshireman, was the co-ordinator at Badajoz, one of the six research centres. This was located in a semi-arid part of Spain where work on forage production under dry conditions was the principal activity. Peter had qualified as a veterinary surgeon in England and after a brief spell in general practice had gone to Cambridge to lecture in clinical pathology. He then took a research fellowship in New Zealand and worked at Ruakura for the Department of Agriculture. The work McMeekan was setting up in Spain was of great interest to Peter and he applied to join the team on a two-year contract in the early seventies. Whilst he would have preferred to have gone to the research centre at La Coruna, in the more humid part of Spain where conditions were more similar to those he had experienced in New Zealand, McMeekan preferred to send him to what proved to be an especially challenging assignment.

The centre at Badajoz, according to Peter, had a brand new laboratory and library, neither of which the staff or students were allowed to use. He recalls that the major problem he found, when he arrived in early 1972, was the Spanish attitude to ex-patriots. They showed great reluctance to accept any advice as to how best to improve their experimental work and as the centre remained under Spanish direction he could exert very little influence.

One important and more promising part of the programme concerned the selection and subsequent training of Spanish postgraduate students. Known as '*Becarios*', these people were to be sent to university or research institutes in Europe, America or Australasia for specialist training. Part of their remit was to identify research needs in Spain before they went, so that on their return, they could make a major contribution in helping to set up the most appropriate work to help solve the problems in their particular field of expertise. This was quite a difficult task as many of them were married with young families and they were reluctant to be away from home for long periods. It was, nevertheless, probably the most successful aspect of the whole project with, in total, over two hundred students gaining invaluable experience overseas.

As time progressed, difficulties within the centres increased even further when the money to purchase equipment failed to appear from the Government. It had been agreed that the World Bank contribution was to be in the form of a pump-priming exercise with significant funds being earmarked from within the country to finance the overall project. At one point Peter remembers that he was asked to authorise payment for some expensive equipment, purchased years earlier and used for peach tree spraying and harvesting. This had nothing to do with the Bank project and it seemed that the staff in the centre were prepared to go to any lengths to clear their accumulated debts. There were examples from other centres where relationships between the research co-ordinators and the Spanish personnel pointed to an unworkable stalemate. McLaughlin in his book refers to a remark made by John Gerring concerning one such instance:

Mac [McMeekan] had got a man over there, a New Zealander by birth actually but just by coincidence, who was from Riverside, California. He was the recognised world authority on citrus culture. There was no question about that. In fact he was so highly regarded that when they organised an international research conference they held it in Valencia because this man was there at the time. Well, the Spanish made this man sit on his arse, near enough for two years before they would even talk to him. They just couldn't believe that anybody could teach them anything about these things. It was that kind of problem. They put a director into the organisation that was directly involved in the project locally and he spent two years getting his staff to come up with what he called organocrams. He was an organiser gone mad. He knew nothing about getting on with the job. So in the end the whole thing was wound up, cancelled and the Spanish have since regretted it. They've come back to the Bank and the Bank have said, No.

One might draw the conclusion from these opinions that the whole project was an abject failure. To some, that is probably true, although Peter Mortimor was more generous. In his view, he felt that there were some significant gains and it was unreasonable to expect to suddenly alter a hierarchical system which had been in operation for generations. For example, the assessment of the priority areas for research work was an undeniable step forward. The project also highlighted the need for dedication by those who were working within it. Many of the research staff had previously held part-time jobs that only occupied them in the afternoons. McMeekan, followed by Cooper, insisted that only full time staff were employed. And certainly the *Becarios* were a real long term benefit for agricultural research in Spain.

Aiden Conway, a respected Irish research worker who was the co-ordinator at La Coruna, the centre that researched forage farming under temperate conditions, also took a positive view. He returned to Spain briefly on a consultant project in the mid-nineties and considered that the long-term impact of the World Bank project had been phenomenal. Conway had been more fortunate than Mortimor in his posting and had managed to take his Director on a tour of research stations in the UK and Ireland so that he could see for himself the potential of forage in relation to animal production. On his return, they worked well together and the project moved ahead. He too was especially positive about the success of the overseas training programme for the *Becarios*. He recalls that one of these had been sent to California to work on the problem that had decimated the Spanish citrus industry. Subsequently he had helped to develop a methodology that eventually led to the production of virus-free plants. This was only one example and Aiden Conway had found, on his recent trip to Spain, a high proportion of these people in positions of responsibility and playing an increasingly important role in the country's agricultural research activity. In his opinion, people, rather than bricks and mortar, have been the real legacies of the World Bank project.

Whatever the retrospective opinions of those involved at the time, there can be no doubt that for Mac Cooper, the problems he faced towards the end of 1974 were becoming insurmountable. McMeekan's death had removed the key link between the Bank and the Spanish civil servants. There had been yet another change of President at INIA, who spoke no English and had little experience of agricultural matters. Furthermore he appeared to be an autocratic man with little sympathy for either the people or the problems which had become his responsibility. As a result, communications between Mac, the Bank and the Spanish authorities in Madrid deteriorated as he became wedged irrevocably between the two bureaucracies, neither of which seemed to recognise the difficulties, let alone come to his aid in helping to resolve them.

John Gerring maintained that it would be untrue to say that Mac was fired, more that they found a reason not to renew his contract as he was approaching retirement age, notwithstanding the fact that he had originally agreed to a five-year period. The shock to both Mac and Hilary was traumatic. They received the news shortly after spending Christmas 1974 with the Cravens in Cheshire. Hilary wrote to Barbara on 10 January, a few days after their return to Madrid.

This isn't going to be the easiest letter to write, but I'll try and put it as objectively as I can. We arrived back here at 1 am on Tuesday morning; both feeling pretty exhausted after our thirteen-hour stint of travelling. Mac went off to the office at 9 am and rang to say Dick Powell, the accountant, wanted him to lunch with him. All day I had a feeling of uneasiness. He came back straight after lunch and I don't think I've ever seen him so utterly shattered. Dick had been given the unpleasant task of bringing him a letter – in short, to say he was relieved of his post of Chief Research Co-ordinator and to assume an assignment for the President of INIA to assess and evaluate the agricultural enterprises in the Canary Islands.

The letter went on at some length to explain the reasons Mac and Hilary felt were behind the decision. Principally Mac was not prepared to compromise his integrity and deviate away from what he saw as his job in keeping the Spaniards to the terms of the Agreement. Hilary urged Barbara: 'Don't worry, adversity always brings Mac and me close and he's made a great many good Spanish friends, who admire him tremendously.' Mac added his own postscript to the letter to his eldest daughter.

> I don't think I could have endured the earlier part of this week if your mother had not come back with me. Ironically we returned strengthened with the wonderful family experience of Christmas and New Year and a feeling that our life in Spain was moving to a more even and satisfying keel. As you would expect, because I suffered the first blow, I was the one at the lowest ebb but she, through her sympathy and understanding, and very real courage, brought back a perspective. I am no longer feeling sorry for myself. She has reminded me of the things that matter in life. It won't be long before I am back in Britain being a perfect nuisance to those in establishment with a responsibility for agricultural policy.

It was clearly a huge blow to his confidence and when he had recovered his breath, he wrote a further letter to Barbara later in the month.

> It was not the blow so much, as the way it was given that hurt me. It could have been done so much more nicely. Probably the biggest shock has been to my pride. However I have been allowed to tender my resignation, but I cannot escape the feeling that I have been dismissed. Not because of any incompetence, but because I had been an awkward old b***** in that I had tried to put a brake on irrational spending in order to keep the project on the rails. I had always refused to treat the President of INIA as God Almighty which is a cardinal sin in Spain, which is organised as a pyramid of dictators, each subservient to the one above, but treating those below as though they did not matter. The official excuse is that I will be reaching my retirement age in August and because the project is being extended until 1977, they wanted someone who could see it out. This makes good sense in many ways, especially with a number of new foreign experts arriving over the next four months and the creation of a virtually new team.
>
> I have been given an interesting assignment, partly I think as a sop, but one which I will enjoy. This will last to August or maybe December, which suits me well in respect of getting our home back and making long term plans for our future.
>
> I am not downhearted and in retrospect, I have no regrets and surprisingly little bitterness. I realise I have too many blessings still in my life, principally in my beloved family.

But for all the brave face, it depressed him deeply. Whilst he did some more consultant work after he returned to Britain and wrote articles for the local press, he had lost much of his public platform. Things had moved on. Maurice Bichard, a former student and colleague, who knew Mac well at Newcastle, reflects:

I always felt that his decision to accept a position in Spain was a total mistake. It seemed to have been partly motivated by the need to prepare (financially) for retirement, since his modest university salary, family commitments and earlier lack of planning had left him quite unready. I believe his new post proved much less effective and satisfying than he had hoped, perhaps because he had left all his contacts behind him in the UK. Not only that, but his resignation upset many in the university who had only recently appointed him as a Pro-Vice-Chancellor, and even a few years out of the UK scene meant that his network quickly withered away. Hence instead of enjoying a gradual wind-down in retirement, honoured as an elder statesman and able to ease himself out of jobs at his own pace, when he returned to Britain, he was already largely forgotten. This was a tragedy for someone who had contributed uniquely to the development of UK agriculture after the war and who had stood head and shoulders above all other professors during those twenty years. Challenging times brought forth a worthy leader.

11

'LAST SCENE OF ALL'

Shakespeare's 'Seven Ages of Man' describes the
   'last scene of all,
   That ends this strange eventful history,
   Is second childishness and mere oblivion,
   Sans teeth, sans eyes, sans taste, sans everything.'

Sadly, such a description is for some people appropriate, and the burden is a heavy one to bear for their nearest and dearest. Fortunately in Mac Cooper's case, his retirement years, for the most part, were active and productive. Although towards the end of his life his physical health did decline markedly, he never lost his mental capacity and, just as important, his determination to live life to the full.

After their return from Spain and the completion of the 'sop' assignment in the Canary Islands, Mac and Hilary thought long and hard about their future. There never had been any serious question of returning to New Zealand. Even if Mac had wanted to, he knew that it was a non-starter as far as Hilary was concerned. And in any case, the proximity of their rapidly expanding family meant more to both of them than anything else. They had also come to love Northumberland with its open sky, rolling hills and wonderful coastline. Perhaps more important than anything else was the fact that they had many friends in the area who had welcomed them back with warmth and sympathy. This helped Mac especially to put the Spanish experience behind him. They were truly happy to be adopted 'Geordies'.

Whilst they still had the tenancy of Holme House, it was much too big for the two of them and it seemed sensible to buy something of their own. At least one benefit from the Spanish project had been a substantial boost to their financial circumstances, so they had the time and the cash to look for a property which suited their needs. Not for them the tidy suburban housing estate, with a modern bungalow and easily accessible shops. They wanted something with character from which they could create their own 'Cosy Ship'.

The opportunity arose in June 1976 when they purchased a dilapidated terraced cottage in the nearby village of Longhaughton. It needed complete renovation. During the months of rebuilding they continued to live at Holme House and could supervise the work as well as make a start on laying out a new garden. It was an exciting challenge at last to have the opportunity of planning a home of their own and to be able to work on it together. It was not until the end of 1977 that they eventually took up residence, and in Mac's customary Christmas letter to Thelma and Joe, it was clear that all the hard work and investment had been worthwhile. Mac wrote from their new address at 'Holme Cottage'.

Our new house is the cosiest we have ever had. We are absolutely thrilled with it and any misgivings we might have had about leaving Holme House for a dwelling but a third of its size have not been realised. The star feature of the cottage is our lounge, which was formerly a workshop with a dirt floor and rough beams supporting a pantile roof. It had rough stone walls and a garage type door at one end. An attractive stone fireplace has now replaced this, the beams have been cleaned up and the walls have been plastered. We also put in a bay window which takes advantage of any afternoon sun that is going and it adds considerably to the character of the room. In fact it is more like a sun porch than a bay window and it gives a measure of spaciousness to the room and this is important because it has to serve as dining room as well as lounge.

Mac and Hilary in their beloved garden at Holme Cottage.

185

Mac then went on to describe the rest of the house in some detail, with the only regret that he felt the bath wasn't quite long enough to fully stretch his legs. However it was more economical, so there were no real grounds for complaint. The garden, which was to become an important part of their future lives, had to be established on the site of what had previously been the farmyard of a smallholding. Hours of hard labour removing stones and rubbish, followed by the building of rockeries, the sowing of lawns and the planting of shrubberies and rose beds, constituted a formidable task for the two of them, and anyone else they could rope in to help. It was all very different from the years of professional work that had preoccupied Mac for the last thirty years, often spending weeks or even months away from home. This was something that they could plan and build together. Mac, after his spell in Spain, was fitter than he had been for years and had even given up smoking. The whole project was stimulating and hugely satisfying and it brought them closer than they had ever been before.

But manual work, however enjoyable, was not going to be sufficient to occupy Mac's mind. He was not inclined to become involved again with the Agricultural Faculty at Newcastle, not because he expected there to be any animosity towards him, but rather that it was a chapter in his life which was finally over. It was never wise to revisit the scene of former glories. There were two possibilities: the first to undertake some work as a consultant, both in the UK and overseas, and secondly to do some writing.

Following his work with the World Bank and the Food and Agriculture Organisation (FAO), he had built up a long list of contacts. Although the Spanish project had gone sour, those in the know were under no

Mac and Hilary with eldest daughter Barbara in the front garden they created at Holme Cottage, Longhoughton.

misapprehension as to where the fault had lain. Mac had had to carry the can for a failure that should have been foreseen by those at the Bank headquarters. His reputation was still intact. In 1976 there arose an offer to review the agricultural development potential of the major islands in the Caribbean. This was a substantial assignment and for some of the time he had the bonus of Hilary being able to join him on the trip. The organisation of this particular project was somewhat haphazard owing to the chronic communication problems in that part of the world, coupled with the fact that the West Indian sense of urgency tends to fade quite quickly as the day wears on. He completed his report, although he felt that it had probably been consigned to gather dust on the shelf of some politician with a vested interest, rather than, as he had hoped, become a working document from which to formulate a coherent strategy for the region. There arose, however, another interesting opportunity on the mainland of Central America, in the former colony of British Honduras, now renamed Belize.

This Commonwealth sponsored project, drawn up by economists in the Caribbean Economic Community's headquarters in Georgetown, Guyana, appeared to them to be a modest and straightforward task. The aim was to evaluate the practical aspects of establishing a complex of dairy farms on 30,000 acres of land, which had been cleared of valuable mahogany and had reverted back to thick jungle. It was an ambitious concept with the combined objectives of providing a much needed source of fresh milk, a working example of the potential of modern farming techniques, and an opportunity for local people to gain employment as well as education.

Following on from Mac's experience of his previous work in the Caribbean, he was asked to recruit a group of experts to visit Belize and carry out a full feasibility study. His team was a multi-national one and included former colleagues who had experience in land development, agricultural mechanisation, farm management, dairy processing and agricultural education. The six-week visit in the summer of 1977 coincided with a flare-up of political unrest, as neighbouring Guatemala was threatening to invade the country to reclaim land which they argued was rightly theirs. Thus the presence of British troops and military aircraft roaring overhead were something of a distraction to the team as they established themselves at their headquarters close to the disputed border. Their unease increased somewhat following a visit by the local commanding officer on the first evening of their arrival. He told Mac to ensure that his staff were on the right side of the river if the balloon did go up, as the first thing the army would do would be to blow up the bridge!

As so often in those days of overseas development projects, it quickly became clear that the practical problems of establishing an enterprise of this size from scratch were immense. Writing some time later in a local Northumbrian newspaper about the experience, Mac outlined some of these problems. There was no infrastructure such as road access, electricity or water supply. Whilst there were ways and means of rectifying the shortfall in these requirements, it had not occurred to the desk-bound economists that there would be enormous on-costs to provide these basics. Neither was there any experience or expertise in the country (which is about the size of Wales) with regard to specialised milk production. There was only one milking machine in the whole of Belize, this having been donated by a charitable organisation in England. It was not however in use on the Government farm where it had been installed, as the staff preferred to milk the small dairy herd of Brown Swiss cows by hand! It was self-evident that ex-patriot staff would have to be employed in considerable numbers if the project was ever to get off the ground and this alone added a substantial extra cost to the project.

Finally there were numerous technical and logistical problems, not least of which was the available source of dairy cows. It was possible to import temperate breeds such as the Holstein, but their susceptibility to heat, tropical disease and indifferent management made them a more risky option than grading up or cross-breeding, using the indigenous cattle population. Certainly it was a much cheaper alternative, but progress towards a productive dairy herd would, inevitably, take much longer. The result of the study was that the team felt unable to support the original proposal of a rapid build up to a complex of 250 farms and 20,000 dairy cows on the identified site. This would have involved an investment, covering production and processing, of well over £10 million, which would have to be funded by overseas borrowing. The advice was to make haste much more slowly and to create a small nucleus herd that would not only serve as a springboard for further expansion, but would also establish the production parameters which are an essential base from which to plan future progress. Whilst eminently sensible, this was not the advice the economists and civil servants wanted to hear. They urged reconsideration. But Mac was adamant that he would not recommend action that was unlikely to succeed in practice. There were far too many examples of over-ambitious projects funded by overseas debt that were not only a disaster, but added a considerable financial burden to developing countries who could ill afford it.

Mac had always felt a responsibility to try and help those in less developed countries to help themselves by improving their agricultural production. He felt strongly that direct food aid was a short-term fire-fighting tactic that did not address the basic problem of a lack of understanding as how best to utilise the available resources. It was far too easy for consultants to come in and rubber-stamp grandiose schemes costing millions, and then leave the unfortunate troops on the ground to get on with it. He was not going to be a part of something he did not believe would work.

Despite his wish to contribute, he therefore left overseas consultancy work, with some disillusionment and with a feeling that his endeavours, both in Spain and in the Caribbean, had been unsuccessful. Had he been more amenable to the wheeling and dealing that surrounded many of these so-called development projects, maybe he would have had more to show for his efforts. But he had non-negotiable standards and he would not be compromised. Certainly it seemed to him that there was much too much evidence of badly thought out strategies and misdirected funds placed in the hands of incompetent administrators. Perhaps also, at the age of sixty-seven, he felt he was getting too old to continue travelling round the world trying to help solve problems of such magnitude.

In his article about Belize, he concluded by saying: 'We may think we have problems in our farming but today the real challenge in agriculture is not found in the sophisticated countries of the western world, but rather in countries like Belize, which need guidance, capital and above all, the confidence and determination to use the natural resources of the country to raise standards of living.'

Back in the UK, Mac dabbled in some contract consultant work for a firm of national land agents. No doubt they saw some merit in advertising the fact that they had Mac Cooper on the books. One of his more pleasurable commissions was to visit and advise Phillip Merricks about how to overcome problems of unthrifty lambs on one of his farms in Kent. Phillip was the son of an old friend of his when the Coopers lived in Kent, some thirty years earlier. Mac's only regret in visiting him was that his employers would want their 'pound of flesh'. But the truth of the matter was that farming in the UK had moved on yet again during the few years he had been overseas. In addition, clients required financial forecasts and cash flow budgets for their bank managers, rather than sound husbandry advice on livestock production or grassland management. Money matters were not Mac's scene and besides, he wasn't comfortable having to charge fees for advice, which he preferred to give freely.

Although still in demand for the occasional talk to farmers' meetings, Mac no longer felt the urge to provoke the politicians, the Ministry of Agriculture or breed society committee members into action. His offerings were therefore less controversial and newsworthy. One event which he did attend for five consecutive years, was the Dairy Farmers Conference, jointly staged by the Welsh Agricultural College and the Milk Marketing Board's Consulting Officer Service. The fact that Dai Morris, by then Principal of the College, and John Craven, as head of the MMB service, were both sons-in-law, may have been one reason why he agreed to give the summing-up on each occasion. Nevertheless it was evident from the audience appreciation that he had not lost the ability to put his finger on the key issues and to send farmers away thinking very carefully about how they might apply some of the ideas that had been exchanged.

It was to the written word that Mac turned, as his last major contribution on matters agricultural, during his last decade. He had always enjoyed writing, having a natural flair and an uncomplicated style that made for easily digestible reading, even for farmers after a hard day's work. For a while in the late 70s and early 80s he wrote a regular article for the *Advertiser*, a weekly newspaper publication with a circulation throughout Northumberland. The topics tended to be rather historical and retrospective, and for the most part dealt with some of the more successful trial work that had been carried out at Cockle Park in the sixties.

In one of these, dealing with the work that Bob Thomas and Murray Black had undertaken in the control of nematodirus infestation in sheep, it became clear that his message was rather critical of farmers. Rather than adopting the less costly 'clean grazing technique' that interrupted the life-cycle of the parasite, the tendency was to rely increasingly on the agro-chemical and pharmaceutical companies to provide drenches and drugs to combat the effects of infestation once it had become established. Mac felt it was no surprise that cure was becoming more popular than prevention because there were large advertising budgets and hundreds of company representatives out on farms, extolling the virtues of the most recent 'wonder drug' on the market, whatever it might be. Maybe farmers ought to think more carefully about the consequences of adopting such a false sense of priorities and reconsider taking action to put prevention ahead of cure, especially when there were proven systems to follow.

As a New Zealander, Mac had always had serious doubts about Britain's entry into the Common Market and he wrote an article in 1980 on the subject, which appears not to have been published. This was at a time when the cost of surpluses was threatening to bankrupt the Common Agriculture Policy and before the eventual introduction of production quotas, co-responsibility levies on cereals and the iniquitous 'set-aside' scheme, where farmers were paid not to plant crops. Twenty years on, many of the problems appear just as intractable.

> My anti-Market feeling is by no means new. As a New Zealander, resident in Britain, I have long had a strong belief in the Commonwealth and its institutions. And back in those days when British entry was debated, sentimental reasons apart, it seemed to make more sense, on both economic and political grounds, to enlarge rather than restrict trade between member nations. The prospect in Europe seemed to be one of taking in each other's washing but in Commonwealth countries there was reciprocity in production and demand patterns.

He went on to remind his readers about the contribution that Commonwealth countries had made to Britain's cause in time of war, and the relief which many felt when de Gaulle had said '*Non*' to Britain's application to join. But the politicians were determined to get in on any terms and now what was the result?

Our lives are being affected to an extraordinary degree by a faceless bureaucracy in Brussels, taking decisions on such diverse matters as 'the spy in a lorry cab' and the amount of meat in a pork sausage. We have only been able to keep our Milk Marketing Boards by making concessions in other directions that can only result in the enlargement of the unwanted food surpluses. On top of this Britain, the poorest country in the Nine except for Eire and Italy, is now the largest net contributor to the Community's purse. The biggest demand on that purse comes from the Common Agricultural Policy and this, looked at dispassionately through British eyes, cannot but be described as an unmitigated disaster.

The article goes on to berate the state of the European milk market and the undisputed fact that Britain's national herd size would have to reduce in spite of the fact that she was not self-sufficient in this commodity. This was apparently to provide a market to accommodate the small, highly subsidised, and inefficient continental producers. Little did he know then, that within four years, Britain's dairy farmers would have milk quotas imposed upon them which effectively handed the shortfall in their own home market to their competitors. Sugar was another commodity that prompted a paragraph in the article.

Sugar provides another example of over-production in relation to the demand that can be generated with high price levels. Again it is being dumped on the world's markets and this is at the expense of Mauritius, Fiji, Guyana and a number of other small countries where the economy is highly dependent on this crop. Some of these are old slave colonies of our shady imperial past for which, in justice, we still carry a responsibility. There is something obscene in my view, in a situation where a wealthy bloc like the EEC selfishly destroys the livelihood of other countries who can, in fair competition (as New Zealand is able to with butter, cheese and lamb) give the consumer the benefits of more efficient production.

He concluded the article by recommending that farmers examine their costs, not just the higher prices they had enjoyed in recent years. They couldn't blame higher oil costs on the EC, but unquestionably, the cost of labour and general inflation had increased as a direct result of higher food costs within the Community.

There is an old peasant's saying in Denmark to the effect that 'when you wet your pants, the feeling of warmth is very temporary'. The initial feeling of prosperity following entry is waning as every cost, from rents, fertilisers, interest charges and repairs to farm machinery rises, and in moving from a low cost to a high cost structure, many in farming are starting to feel the cold. Unfortunately it is now too late to unscramble the egg and we will have to grin and bear the doubtful privilege of very expensive membership, unless our 'Iron Lady' is able to do something to change the situation that her predecessor, as Conservative leader, helped to create.

The early eighties saw the publication of updated editions of Mac's successful books on grassland farming, sheep and beef production. Written in conjunction with former colleagues at Newcastle, these works were still in considerable demand from students and farmers alike. But he also wanted to try something new. Encouraged by Hilary, he started his autobiography and his initial efforts form part of this book. But he disliked talking about himself and tired of the idea before the story reached the more influential part of his life.

He even tried his hand at writing a novel, and wrote to sister Thelma to describe the plot. It was to be a story about a young Scot who after leaving college goes out to New Zealand to get farming experience. He subsequently marries his boss's daughter, who with their two-year-old child dies in a motor accident. He returns to a run-down farm in Northumberland and becomes involved with a Roman Catholic woman who is separated from her Irish husband, whom Mac describes as 'a proper little bastard'. No record exists of draft manuscript, but the letter to Thelma asked numerous questions about the times of various events which took place in Napier in the 1950s, in order to authenticate the passage he was then writing about his hero's train journey from Wellington. It seems, however, that this project also failed to progress beyond the idea stage as the subject was not raised again in future correspondence.

One of Mac Cooper's last major contributions was to write some 30,000 words for the sixteenth edition of Fream's *Agriculture*. Originally published in 1892, this book ambitiously attempted to describe all the main elements of the farming industry. The 1983 edition, edited by Colin Spedding and prepared under the authority of the Royal Agricultural Society of England, does not attempt simply to update previous publications. Rather it takes a completely new approach, concentrating on the main production systems in agriculture and the principles that lie behind them. Mac's input came in Part One of this formidable tome of over eight hundred pages. In what was a scene-setting section, he outlined the structure of agriculture from World, European and British viewpoints, detailing the major changes and developments which had taken place in recent times. Whilst he admitted to finding the task a hard one, the result is a well balanced overview of a remarkable success story, if judged by the technical advances in farming practice and the consequent increases in productivity of both crops and animals. But the subject matter for him was rather bland and lacking in opportunity to provoke his readers into thinking for themselves with regard to the future of farming in the developed as well as the developing world. About this he did have strong opinions, but a reference book of this nature was not the medium through which to express them.

One remaining honour, which Mac was delighted to receive in June 1979, was to be appointed a Fellow of Wye College. Back in 1949 he had introduced Sir Edward Hardy for this award at the annual Commemoration Ceremony, amidst a background of academic sobriety and student high-jinx. It was Bill Holmes, Mac's successor as Professor of Agriculture, who had the job of introducing him on this occasion. The other Fellow to be so honoured was a Professor Selman, introduced by the 'Welsh Wizard', Professor Wibberly. According to the Principal's report on the proceedings: 'An already memorable occasion, was made even more so when Professor Wibberly broke into song – stimulated perhaps by a combination of Welsh temperament, brilliant sunshine and the sheer pleasure of the day.'

It proved to be a splendid reunion in that it allowed both Mac and Hilary to meet many old friends and former colleagues. One of these, Donald Sykes, remembers Mac giving a talk to the Daniel Hall Club (dealing with student affairs) at Withersdane Hall during the weekend's festivities. According to Donald, Mac reminisced about his time at Wye, and became surprisingly emotional, with tears flowing freely, as he recalled the most enjoyable years of their lives. Maybe, in perspective, Wye had meant much more to the Coopers than it seemed at the time.

Maurice Bichard relates another example of this emotional dimension.

My last clear memory was of him travelling down from Northumberland to be present at the Café Royal when I was presented with the David Black Award [for outstanding contribution to the British pig industry]. At the end of the lunch, when I had inevitably identified him as one of the three major influences on my life, he was unashamedly in tears, so moved by this recognition of, and by, one of his young men.

In September 1981, now having passed the allotted span of three score years and ten, Mac decided he wanted to return to Italy. This was to show Hilary, as he put it in a letter to Thelma, 'where I fought so gallantly for King and Country as well as liberating a lot of local wine.' He particularly wanted to visit Ischia, an island near Capri in the Bay of Naples where 'I spent a week in 1944 just after we pulled out of Cassino, enjoying hot baths and sleeping between sheets.' He didn't mention returning to Cassino itself; perhaps the memories were too painful.

The following year, Mac and Hilary made their final trip together to New Zealand. They went in January, leaving behind a freezing English winter. In a letter to Thelma anticipating their trip, Hilary longs for the New Zealand sunshine. However, she explains that they are well insulated from the worst ravages of the weather in their little cottage.

How good it is to live here, because really there is no necessity to go out at all. Our butcher, baker, fish girl, etc. plus the deep freeze make shopping unnecessary. Also this week we have been treated very well as OAPs as first came a food hamper from the RAF (they have a station just outside the village) and then today we both received envelopes containing a £5 note from the Longhoughton Senior Citizens' Committee. I feel slightly embarrassed by it all and last year gave my £5 to the church, but after doing all my Christmas shopping, have decided that charity begins at home.

The trip was a great success, as this time Mac was spared the distraction of a full itinerary of speaking engagements. It really was a chance to meet all their friends and to visit places in both islands that neither of them had been to before. Mac describes, in a letter to his eldest daughter Barbara, one sentimental journey they made to Waimaramara.

I always loved this place, with its golden sands and Pacific rollers and it was there that we spent the first fortnight of my return to civilian life. It was the same sort of weather then as we have been enjoying since our arrival. You and the twins were waiting for us at a bridge, dressed in brown check frocks and you were all as brown as berries. I got out of the car and walked with you up to the cottage that your mother had taken. It was one of those occasions when I needed three hands.

On returning home, Mac wrote to Thelma and Joe to express their gratitude 'for a truly wonderful trip'. They had been to parts of New Zealand that they had only read about, such as the Coramandel Peninsula and the State Forest in Northland to see the massive kauri trees. He wrote: 'There are still parts such as the west of Gisborne which remain for our next visit, but I am beginning to feel my age and we will have a stop-over on route, in future.'

There was not to be a next visit for Mac and they probably both secretly knew it as they returned, in beautiful Easter weather, to their cottage in Northumberland. They were thrilled that Mattie and Betty, the local couple who devotedly helped in the house and garden, had been busy decorating the house and Mac's letter was full of the emerging spring bulbs and his activities in the greenhouse. It is clear that they were delighted to be home.

One of Mac's favourite activities centred on the vegetable garden. They had managed to acquire an additional strip of land from the next door neighbour so that there was plenty of opportunity for growing potatoes, carrots, onions, spinach, lettuce and, most important of all, leeks. The growing and showing of leeks has been a popular pastime for generations of gardeners in the North-east of England. This is true particularly in the villages bordering the traditional coal mining areas, where many of the competitors lived in rows of small terraced houses without access to their own garden. In these cases, the leeks would be grown, along with a wide variety of other vegetables, in Council owned land, subdivided into small individual 'allotments', and available on payment of a peppercorn rent. Irrespective of where or how the leeks were grown, all the competitors had to belong to the local club and abide by the rules, which were rigorously enforced. During the year everyone was compelled to make a weekly contribution towards the prizes as well as to the ensuing evening's entertainment. Most people ended up with a prize of some sort, but at the top of the list there were substantial awards which were much sought after. Needless to say the competition was extremely fierce and much local kudos was attached to those who achieved success. Mac's first reference to his leek growing aspirations appeared in a letter to Thelma, shortly after they moved into the cottage.

My leeks, one of 35 entries were awarded fourth prize, which consisted of a very fine clock in a glass case. The price tag reads £49.50 so my modest efforts have been well rewarded. When the stewards came round to stamp my leeks (as a safeguard to any duplication) they said I should do well, but I did not dream that an old-timer like me should be so well placed. My triumph was shared by Betty and Mattie. Mattie selected the three leeks that were entered and he prepared them for exhibition. The final touch, after thorough washing, was a brushing with milk to give the leaves and roots sheen.

Expert leek grower.

Mac refers to the subject again in his letter to Thelma upon returning home from their latest trip to New Zealand.

> Most important of all, our respective leek beds have been dug over and fertilised. I have about 40 leek plants now in pots, which will be fairly allocated between husband and wife. I am going to have to be on my mettle this year, because it would be a great blow to my pride if Hilary, an absolute novice, were to be higher placed than me. At the same time I cannot take any unfair advantage over her, so she has first choice of leek beds and we will take turn about in selecting the individual leeks for planting.

In fact neither of them did well in the following autumn. Mac fell from grace to eighteenth, beating Hilary by one place, much to his relief. Even so, they came away with electric toasters, tableware and glasses, all earmarked as Christmas presents for various members of the family.

Another activity that they both enjoyed was playing golf. Mac might have been a star on the rugby field in his day and a respectable cricketer and tennis player to boot, but golf was not on his list of sporting achievement. His swing might be best described as agricultural in that what it lacked in style appeared to be generously compensated for by physical effort. He usually hit the ball hard, but more often than not, neither he nor anyone else saw where it went. Hilary on the other hand was a more talented player, which occasionally rather irritated Mac. To be beaten by a woman at anything was to him something of an anathema.

John Craigs and his wife Liz, both golfers, and who had been farming neighbours and close friends of the Coopers' when they lived at Tritlington, recall some raised eyebrows at their golf club at Foxton Hall, Alnmouth, shortly before Mac and Hilary became members. Apparently Mac and Hilary had been putting out on the ninth green and well within earshot of both the clubhouse and the first tee, when a burst of kiwi invective shattered the stillness of a summer evening. The word went round that it was Professor Cooper and he must have had a sudden lapse of memory as to his whereabouts. Nevertheless they duly became members and enjoyed many happy years of companionship on the course and at social events at one of the best golf clubs in the North-east. In later years when Mac could no longer play owing to ill health, Hilary would often be seen playing a round followed by their faithful Labrador. Cara had been trained to find lost balls and on emerging with a find from the deep undergrowth, would be rewarded with a piece of chocolate. Everyone in the family who played golf on holiday in Northumberland always left with many more golf balls than they brought with them.

In fact, dogs were important members of the Cooper household ever since the early days at Wye with Kiwi, their golden retriever. There followed a pair of Bedlington terriers in Northumberland, called Peter and Susie. They are an unusual looking breed, resembling something between a new born lamb and a grey poodle. They then chose Labradors, the first being a black called Teresita, followed finally by Cara, the golden, golf ball finding, bitch. All were much loved and hopelessly spoilt, more often than not occupying pride of place in front of the fire on winter evenings. And as with all dogs, they relished the long walks through the countryside or along the miles of beach with Hilary, whatever the weather.

Whilst Mac loved his daughters dearly, it was not until they approached adulthood that he began to develop a close relationship with them. His idea of family holidays had never amounted to much more than a few days sandwiched between work commitments. In fact everyone tended to heave a sigh of relief when he had to rush off to an important meeting, especially if it was in the middle of a rain-soaked week spent in a caravan. Maybe he would have been more participative had he had a son, especially if he had a flair for rugby. But he had a low boredom threshold and lacked the patience to hold the attention of the very young. He was constantly preoccupied with his own thoughts surrounding his work.

His daughter Diana writes of her father:

Although his word was law and extremely strict at times as well as rather a remote figure, I had enormous pride and respect for all he achieved. To listen to him on the radio, watch him on TV, see his name as author of books and to hear people say in awe – is Prof. Cooper your father? All this helped to give me an inner confidence and ability to cope with my future life. My mother gave us the everyday love, care and joys of a very happy childhood which balanced remarkably well with a famous and much in public demand father.

As his nine grandchildren began to grow up and to visit for holidays, Mac preferred to remain somewhat apart from the rough and tumble. Occasionally their boisterous behaviour would goad him into action and they all knew that he was not to be trifled with. A number have cause to regret their actions such as watering his prize tomatoes with kerosene or showering the passers by with soil from the window box of their seventh floor apartment in Madrid. In their youth, the grandchildren were wary of 'Grandpa Prof.'. Indeed he was not averse to commenting in letters to Thelma that in his view, some of them were rather over indulged. He wasn't comfortable with the rapidly changing family values and social mores, he preferred children to be seen rather than heard. By the eighties of course, they were well on the way to adulthood themselves and his relationship with all of them matured and prospered in much the same way

197

as it had done with his daughters.

In fact three of them, Susan Craven, Fiona Thompson and Peter Craven, were to undertake higher education in the North-east, so this provided opportunities for them, as impecunious students, to visit the cottage at Longhoughton for Sunday lunch and at the same time to stock up with food and wine. Fortunately by this time, Mac had begun to make his own wine. A friend of his who had been investigating the feeding value of grape concentrate at Cockle Park had acquired a substantial quantity from a Common Market source, nine gallons of which had come Mac's way. He had taken over the kitchen in the cottage and at any one time had about twenty gallons of the stuff bubbling away at various stages of fermentation. It was more like a laboratory than a kitchen. Recipients of his efforts varied in their opinion as to the quality of the vintage; however for students, it contained alcohol, there was plenty of it and above all it was free.

The girls were engaged in studying biological and business studies, whereas Peter was an agricultural student at Newcastle. But perhaps of more interest to Mac was that Peter appeared to be a useful rugby player. In a letter to Thelma in November 1983, Mac says of Peter: 'He is 6 ft. 2 ins. and just on 14 stone and there is promise of more to come. He is now playing for the University after a not very promising debut. His first game was with the University sixth team and the referee, a woman, was whistle happy. I would have loved to have seen John's face when he heard that item of news, because Peter's rugby matters very much to him. Now expectations are being satisfied as he follows in his father's footsteps.' Peter's involvement in rugby encouraged Mac to go and watch the University side on many occasions over the next three years, re-establishing contact with some of the personalities who had been involved with the game when he had been President of the club some twenty years previously. Unfortunately when Peter graduated three years later, a nasty attack of shingles prevented his grandfather attending the award ceremony.

Mac took a great interest in the careers of his former students, be they graduates of Massey, Wye or Newcastle. Over the years, he must have written literally hundreds of references on behalf of those who were seeking employment. And long after he was active in the academic field, he continued to receive requests to support their applications. Although in the later years he had less opportunity to keep in touch with his old students, there remained one group who had the pleasure of his company when they had their annual reunion. In a letter to Thelma and Joe in October 1984, he explains the background to his involvement.

It so happens that this weekend marks the annual get together of the ancient and honourable Society of Newts, of which I am an honorary member. The hard core consists of John Craven's contemporaries at

Newcastle, mainly rugby players who started their get together twenty-eight years ago. With maturity and a capacity for more rational behaviour, I felt it safe to suggest to them about fifteen years ago that they should base their weekend on our golf club and this together with the Schooner Hotel in Alnmouth have figured ever since. There is a golf competition on the Sunday morning with a number of trophies as rewards for meritorious effort. The golf can scarcely be described as being of outstanding quality but the company is good and the golf club bar does a very good trade.

This group celebrated their forty-fourth reunion at the same golf club in October 2000. Not all of the people involved were former students of Mac's. Indeed the Honorary Secretary, Tom Nicholson, was a classical scholar! A handful did not even attend the University. However they all had the good fortune of being on the Newcastle scene in Mac Cooper's heyday and the friendships they made in those formative years have truly stood the test of time. Mac enjoyed that fellowship, indeed he was very much part of it, and all those who still attend those annual functions remember him with great affection and gratitude.

Mac with his first great-grandson Ben Craven, 1988.

In April 1986, the first marriage of the grandchildren's generation took place in Cheshire, and much to everyone's delight Mac and Hilary were able to attend. At David and Valerie Craven's reception, Mac was asked to toast the health of the bride and groom. In his speech he pointed out that it was ninety years to the day since his mother and father had married and what a happy marriage that had been. He added that they had had seven children and if the guests thought that was a little excessive, he reminded them that he was number seven and there would have been no wedding and no bridegroom if his had been a family of present day dimensions. He had lost none of his touch in speaking from the heart. Sadly this was the last family wedding Mac was able to attend, as his health started to deteriorate and travelling away from home became too stressful. It was fortunate, however, that Ben Craven, David and Valerie's eldest son, appeared on the scene at the end of 1987, so making Mac and Hilary great grandparents for the first time.

Their extended family had not only grown up with breathtaking speed, but their daughters' lives and circumstances had inevitably changed as well.

The Morrises had for some time been dairy and sheep farming in South Wales. Dai had tired of the politics of university life, having successfully established the Welsh Agricultural College in Aberystwyth, and he was determined to pursue his first love of practical farming. Cynthia played her full part on the farm and her skill with animals seemed to come as second nature. Indeed her father was reputed to have said on one occasion, that the welfare of the calves appeared to come before that of either her husband or her children. Sally and David, still at home at this stage, looked set to stay in the area and work within the agricultural community.

Diana's home life had been less fortunate. She had embarked on a teaching career at primary school level, eventually achieving the post of deputy head, but sadly her marriage was to break up. She did, however, soon meet Mike Stevenson, who also lived in Lincolnshire, and eventually married him. Diana's three children, Michael, Fiona and Philip, were each following different paths, none of which had anything to do with the farming industry.

In the early sixties the Cravens had purchased a run down farm in Cheshire and ran this as a sideline to John's job with the Milk Marketing Board. Barbara, after producing four children in quick succession, recommenced her medical career and joined a general practice in Chester as a partner. Two of the boys, Peter and David, started work in the agricultural industry, but Peter soon branched out into a career in marketing. Robin, the youngest, attended Loughborough College to study sports science and in 1987 was still undecided what to do. Susan, the eldest, married Neville Sorrentino in the spring of the following year and they set up home in North Wales, where Neville had started a business of his own.

On 11 September 1987, Mac and Hilary celebrated their golden wedding. The anniversary was held at Spy Hill farm in Cheshire, the family home of the Craven clan. The event spanned three days and as well as the entire family, numerous old friends of the happy couple were located and joined in the festivities at various points. It was one of those occasions where everything seemed to run like clockwork; even the weather was exceptional. Beneath the surface, however, everyone knew that it didn't just happen, it was the result of months of careful planning, masterminded by Barbara and aided by her twin sisters. When they got home Mac wrote to his eldest daughter:

> The euphoria generated by the events of last weekend is still with us and will remain for a very long time. It was certainly the most memorable event of our lives. You may say, but what about your wedding day, but then we did not have the lovely family that stemmed from our marriage. Friday night, in particular, will be an outstanding memory because all of us were together in an expression of family unity that centred on your mother and me. The rest of the weekend was in keeping with the spirit of the opening night. It was a happy thought of yours to bring so many friends together on Saturday night including the Peto's and Ken Deighton,

Golden Wedding family group at Spy Hill Farm, Cheshire, September 1987.
Standing left to right: John, Robin and Peter Craven, Sally Morris, Mike Stevenson, David Morris,
Fiona and Philip Thompson, David Craven, Dai Morris.
Front: Valerie Craven, Barbara Craven, Hilary and Mac, Diana Stevenson,
Susan Craven, Cynthia Morris.

who were representative of that outstanding bunch of students I first had at Wye.

Our lovely presents will be a reminder of this wonderful occasion and the way our lives have been enriched by it. Thank you again for what you have done for us.

Sadly by now Mac's health had taken a turn for the worse. He had developed acute emphysema, a condition of the lungs causing severe difficulty in breathing. Although he had given up smoking some years earlier, the damage had been done and he was now paying the penalty. For a man who had always been active, and now in retirement enjoying his garden, his condition was a cruel blow. It meant that even the slightest exertion brought on breathing difficulties and this was accompanied by bouts of depression. But Hilary was a tower of strength. Their roles effectively reversed in the latter years of Mac's life. She did the manual work, helped in the garden by the indomitable Mattie and in the house by Betty. Mac pottered in the greenhouse, continued to make copious quantities of wine and also cooked many of their meals. But they could not venture far from home.

Mac's last letter to Thelma and Joe (who had recently suffered a stroke) was dated 19 May 1989. It is full of news about the garden and Hilary's sterling work in planting out all the seedlings, which they had raised in the greenhouse. He ends his letter:

This time of the year always recalls your visit with the rhododendrons, azaleas, laburnums and lilacs in full bloom. Outside my window as I write, our crab apple is making a magnificent show, and this for the very first time. Next to it, our red may is showing a lot of flower buds, so it will soon be making its contribution. The roses are forming buds, especially the 'All Gold' floribundas that the grandchildren gave us.

Wouldn't it be wonderful if one of your beloved daughters would organise a magic carpet to bring you over, with appropriate comfort stops, to share a slice of the contented life we lead these days? Incidentally nobody is more contented than Cara who is the most indulged but also the most lovable of dogs. She is at my feet as I write.

Mac Cooper died some three and a half months later on 1 September 1989. He was seventy-nine. He had been at low ebb for a few days, although his condition had not given cause for undue concern. A sudden worsening led to his doctor calling for an ambulance to take him to the local hospital in Alnwick. With Hilary at his side, he died soon afterwards.

The funeral was held at the little village church at Longhoughton and the congregation squeezed into every available space. It was a moving family service, with the burial on the edge of the churchyard overlooking the rolling

'Last scene of all'

Northumbrian countryside. A fitting final resting-place.

Hilary wrote to her niece Sally (Thelma's eldest daughter) in New Zealand on 17 October and that letter speaks volumes about the funeral and the letters she had received.

Thank you dear for your letters and all that you have done. All along I had wondered how I would cope as Mac was getting so breathless and tired. I realise now how much he had deteriorated during the last few months and should have been prepared, but one never is. Also thank you for putting the notices in the newspapers, with the result that I have had letters from old friends who would probably never have known. Two of the first letters I received stand out. The first, the day after Mac died, came from Peter [Sally's godson] a very caring and thoughtful four pages. It ended with this paragraph: 'I wasn't around in Grandpa's heyday, but I have lost count of the occasions and people I have met, that on discovery of my famous grandfather connection, have acknowledged and acclaimed him. Sometimes I even kept quiet about my heritage, for I feared I wouldn't live up to the great standards that he set. All the same I have been very proud to have been his grandson.' The other was from Bill Hugonin, Agent to the late Duke of Northumberland. 'Why was one so happy with him? Always leaving better for his advice and company. Partly I think, because with him, one had to be one's true self; there could be no pretence, bluff or hypocrisy. All was honest and of good report. And of course with him, as with my dear Duke, all men were equal, and the twinkle in his eye said it all.'

You will know by now that I have decided to take a trip out and see you all after Christmas. It is, I am sure, what Mac would have wished as he had said more than once that would be his dearest wish to see you all and his native New Zealand again.

One can't have been married to a Cooper for fifty-two years without so much of the endearing and at times not so endearing characteristics rubbing off on you. I can still see that slightly disapproving look when I have not done things quite the way he would have.

The service was quite wonderful. Just a farewell to a very much loved man. I was very proud of our family. His six stalwart grandsons carrying him out of the church – no mean feat. His three granddaughters, looking very lovely walking alongside carrying sheaves of flowers. Ba, Di and Squint reading their lessons quite beautifully and then John's Appreciation, which I knew he had taken many hours of thought in composing it. The church was full to overflowing with so many old students, colleagues, farming and other friends. Though sad for them, it was wonderful for Pat and Larry to be there to represent the Coopers.

Larry Cooper was Mac's nephew. He and his wife were over on a trip to the UK, and had planned to stay with Mac and Hilary the day after Mac had died. The text of the Appreciation to which Hilary refers in her letter was as follows:

Each of you and the many thousands of people, who either knew Prof. Cooper well or knew of him by listening to his words or reading his thoughts will have their own personal recollections of him at this time. I would like to share some of my thoughts with you. I don't intend to catalogue his achievements or his honours. They are well documented elsewhere.

His industry was agriculture. How will those who work in that industry remember him? Perhaps three things stand out. First, his outspokenness. His legendary remark in 1949, that he thought British farming was at half cock. It not only shook the complacency out of a dispirited industry, it well illustrated the courage of a man who was prepared to stand up and say what he believed to be true. Secondly, he was a farmer's man. He had the vision to recognise the potential application of science into practice. There was nothing he liked more than kneeling in a field of grass, amongst cattle or sheep, provoking farmers into a discussion about its merits and cost saving potential. Finally, as a communicator, he had no equal. Irrespective of the size of his audience he never used a note. They did not know, however, of the intensive preparation he had put into his contribution and the nervous energy expended on their behalf.

His industry was agriculture, but his profession was teaching. How will his students remember him? They also will recall his forthright views, the practical emphasis he placed on matters theoretical, and his inspirational enthusiasm. He was someone worth listening to; a rare accolade from undergraduates when there were so many other exciting things to do. But they will also remember that he listened to them. He was genuinely interested in their ideas and opinions. Whether in a face to face interview, in a seminar, on the farm, on the rugby field or in the bar, he was always the same. And this endured long after they moved on, often to senior positions in the industry. Above all he managed to strike that fine balance between respect and informality. To us, he wasn't Mac, the Dean, or Cooper – just plain Prof.

As you would expect, Hilary has had hundreds of calls and letters of sympathy, many from his former students. One remarked that after his father, Mac had been the greatest influence on his life. Some of you will agree with this. I know I do.

This then the public man, what of the private man? How will his family remember him? I am told that he was a rather strict parent. It can't have been easy having to adapt to a household of four females after the war. As a grandfather, he was held in healthy respect, if not in awe. But as they grew up and got to know the man, his grandchildren each developed a deep affection for him as the various family stories illustrate. All professors are meant to have their little eccentricities and he was no exception. His acute sense of dress. His unusual skill in driving a car; always in top gear, whatever the speed. His ability to invite guests to stay from the very furthest corners of the earth, without having first warned Hilary that they would arrive. All these and many more illustrate his relationship with people as well as his sense of fun. And they also remember his understanding, wise counsel and practical advice with regard to their careers or personal problems. But above all, they remember his love and the family loyalty which both he and Hilary personified.

Two years ago this weekend, Mac and Hilary celebrated their golden wedding at our home in Cheshire. For me, that weekend encapsulates my memory of Prof. The weather was glorious. We spent most of the time outside in the fresh air. He watched one of his grandsons play rugby against London Scottish. He gave a speech without a note. But above all he had his wife, his three daughters and his nine grandchildren all around him. What an example to all of us of the strength and the support of a loving family – in times then, of great happiness, and now in deep sorrow.

One of the lines of the hymn we have just sung was: 'We blossom and flourish as leaves on a tree.' Not all of us do that, but Prof. did, and I know that the farmers he influenced, the students he taught and the family he loved will be forever in his debt.

EPILOGUE

Hilary did indeed go out to New Zealand in January 1990 and enjoyed an extended visit to all Mac's relatives and their friends of earlier years. But it was not the same and she was glad to get back to the cottage in Longhoughton, where she had decided to remain with her memories.

The family expected Hilary to live to a ripe old age. Her mother and grandmother had lived into their nineties and she had had an aunt who had been over one hundred before she died. Furthermore, Hilary seemed fit and well. She played golf regularly, took her beloved Cara on long walks and spent a lot of time in the garden. She also travelled to South Wales, Cheshire and Lincolnshire to spend time with her ever increasing numbers of great grandchildren. In June 1992 she was in great form at Peter and Carolyn Craven's wedding, but unexpectedly the following month, she suffered a heart attack after playing a round of golf with her friends at Alnmouth. Despite a period in intensive care, she did not recover and died, aged seventy-five, a few days later in Ashington General Hospital. The family was deeply shocked. She now rests, alongside Mac, on the edge of the churchyard at Longhoughton, overlooking the countryside they both loved so much.

Since those days Mac and Hilary's extended family has grown even larger as many of their grandchildren have themselves become parents.

The Morrises retired from full-time farming in 1999 and now live in nearby St Clears with sufficient land and sheep to keep Dai out of mischief. Sally is married to Roderick Hancock, who has his own engineering company near Carmarthen. They have two children, Joshua and Kirsty. David works in the dairy industry for one of the milk buyers and also lives in St Clears with his wife Helen.

Diana Stevenson now lives in Crowle in Lincolnshire with husband Mike. She retired from teaching and has recently been fighting illness. After major surgery, however, all the signs are good for her recovery and a return to normal life. Fiona, her daughter, works in personnel in Lincoln, whereas Michael, her elder son is in insurance. Philip, the younger son, has recently moved out to New Zealand and appears to like it so much that he may well stay. One of the advantages of having a parent with a New Zealand passport is the opportunity for his children to stay there without having to go through the longwinded process of qualifying for work permits.

The Cravens remain at Spy Hill farm. John retired as Chief Executive of Genus at the end of 1996. He now has a number of part-time jobs; the farm, watching rugby and playing bad golf to keep him busy. Barbara retired from her medical partnership at the end of June 2000. They are fortunate that their three sons live within a radius of twenty miles, so they see a lot of their grandchildren. David who works on the Duke of Westminster's estate near Chester has two sons, Ben and Michael. Already they are shaping up to be very promising rugby players. Peter now has his own marketing company in Manchester, and he and Carolyn have a daughter, Joanna, and two sons, Charlie and Harry. Robin, the youngest member of the family, spent six years in New Zealand before returning to the UK in early 1999 to work for a Dutch firm who manufacture and install artificial surfaces for playing fields. He and his wife Janet have two small children, Ashleigh and Jamie. Finally Susan, with her family of four, Marc, Nicola, Caroline and Zoe, now lives in the British Virgin Islands. Since 1997, Neville, her husband, has been managing his father's motor business. As can be imagined, they have many visitors from the UK.

In 1990 the University of Newcastle Agricultural Society and the Faculty of Agriculture launched an appeal to raise funds in order to commemorate the life of Mac Cooper. After consultation with Hilary and other members of the family, it was decided that the money would go towards funding an annual Travel Scholarship, to be administered by the Nuffield Trust. Both staff and students of the Faculty are eligible to apply.

Any surplus which might arise from the sales of this book, after meeting the publication and direct research costs, will be donated to this Travel Scholarship Fund.

BIBLIOGRAPHY

Chapter 1

S.W. Grant, *Havelock North from Village to Borough 1860–1952* (Hawkes Bay Newspapers Ltd., Hastings, New Zealand, 1978).

Gordon McLauchlan, *The Farming of New Zealand* (Australia and New Zealand Book Co PTY Ltd., 1981).

Chapter 3

T.W.H. Brooking, *Massey – Its Early Years* (Massey Alumni Association, Palmerston North Printing Co., 1977).

John Mulgan, *Report on Experience* (Oxford University Press, 1947).

Gordon McLauchlan, *McMeekan. A Biography* (Hodder & Stoughton, Auckland, London & Sydney, 1982).

The Bleat (Publication of the Massey Agricultural College Students Association).

The Rhodes Trust and Rhodes House (Oxuniprint, Oxford, 1996).

Chapter 4

Howard Marshall in collaboration with Lieut. Col. J.P. Jordan, *Oxford and Cambridge – The Story of the University Rugby Match* (Clerke & Cochrane, London, 1951).

M.M. Cooper, 'The Demand for Cheese in the City of Oxford' (*New Zealand Journal of Science and Technology,* March 1937).

Chapter 5

Wellington Rugby Football Union Annual Reports 1938 & 1939.

John Ellis, *Cassino. The Hollow Victory* (Andre Deutsch Ltd., London 1984).

Official War History of New Zealand (Volume devoted to the 22nd Motor Battalion, compiled by Jim Henderson, 1958).

An Encyclopaedia of New Zealand: Wars

Great Battles of World War II (Marshall Cavendish, 1995).

Chapter 6

Marygold Rix-Miller, *Trophy of War* (New Horizon (Transeuros Ltd.), Great Britain, 1983

Stewart Richards *Wye College and its world: a centenary history* (Wye College Press, 1994).
Journal of the Agricola Club (Wye College, Ashford, Kent, 1947).
Cardinal (The Journal of the Wye College Union Society, 1948).

Chapter 8
H. Cecil Pawson, *Cockle Park Farm* (University of Durham Publications, Oxford University Press, London, 1960).
M.McG. Cooper, *Competitive Farming* (Crosby & Lockwood, London, 1956).
C.P. McMeekan, I.L. Campbell, M.M. Cooper, P.G. Stevens & A.H. Ward, *Principles of Animal Production* (Whitcombe & Tombs Ltd., New Zealand, Australia & London).
M.McG. Cooper & R.J. Thomas, *Profitable Sheep Farming* (Farming Press Ltd., Ipswich, 1965).
M.McG. Cooper & M.B. Willis, *Profitable Beef Production* (Farming Press Ltd., Ipswich, 1972).
M.McG. Cooper & David W. Morris, *Grass Farming* (Farming Press Ltd., Ipswich, 1961).

Chapter 9
E.M. Bettenson, *The University of Newcastle upon Tyne, A Historical Introduction: 1834–1971* (University of Newcastle upon Tyne, 1974).
Journal of the Agricultural Society of the University of Newcastle upon Tyne Volumes 10 (1954/5) and 18 (1964).

Chapter 11
Fream, *Agriculture*, edited by Colin Spedding (Royal Agricultural Society of England, 1983).